THE CANALS OF THE
NORTH OF IRELAND

THE CANALS OF THE BRITISH ISLES

EDITED BY CHARLES HADFIELD

1. *British Canals. An illustrated history.* By Charles Hadfield. Revised edition
2. *The Canals of Southern England.* By Charles Hadfield
3. *The Canals of South Wales and the Border.* By Charles Hadfield
4. *The Canals of the North of Ireland.* By Dr W. A. McCutcheon

in preparation

5. *The Canals of the Midlands.* By Charles Hadfield
6. *The Canals of the South of Ireland.* By Ruth Delany
7. *The Canals of Scotland.* By Jean Lindsay

OTHER HISTORIES

Waterways to Stratford (Warwickshire Avon, Stratford-upon-Avon Canal, Stratford & Moreton Tramway). By Charles Hadfield and John Norris
London's Lost Route to the Sea. By Paul Vine

RIVERS

The Severn Bore. By F. W. Rowbotham

in preparation

The Kennet & Avon Canal. By K. R. Clew
The Tamar Valley. A study in industrial and transport history. By F. L. Booker

THE CANALS OF THE NORTH OF IRELAND

by

W. A. McCutcheon

WITH PLATES AND MAPS

REPRINTS OF ECONOMIC CLASSICS

AUGUSTUS M. KELLEY · PUBLISHERS
NEW YORK 1971

Published in the U. S. A. by
AUGUSTUS M. KELLEY, PUBLISHERS
New York, New York

Printed in Great Britain by
Latimer Trend & Co., Ltd., Plymouth
for David & Charles (Publishers) Ltd
39 Strand, Dawlish, Devon

CONTENTS

ILLUSTRATIONS

PLATES

Between pages

Between pages

Between pages

XX. Lower Bann: (*above*) the sand quays at Toomebridge; (*below*) sand barge from Lough Neagh
(formerly owned by Guinness of Dublin),
approaching the first lock at Toomebridge   112–113

FIGURES IN TEXT

Page

CHAPTER I

Introduction

THE system of inland waterways in the north of Ireland used for commercial navigation focused on Lough Neagh, the physical framework of the area being such that the largest freshwater lake in the British Isles, touching five of the six counties of Northern Ireland, was linked to the open sea by lowland corridors in which natural waterways and relatively level terrain suggested and facilitated canal construction.

Thus, from the lough, some fifty feet above sea level, the lower valley of the River Lagan provided a natural routeway to the head of Belfast Lough, where from the mid-seventeenth century an expanding port and manufacturing centre had been developing. However, as the eighteenth-century opened and the concept of inland navigation along artificial cuts emerged as an essential factor in the growth of industry and commerce already beginning to gather momentum throughout many areas of Great Britain, Belfast was not sufficiently important to warrant the construction of a navigation channel linking the eastern seaboard with Lough Neagh, and it was in the south-east, towards Newry, that canal construction began.

Here, a lowland corridor between the uplands of Down and Armagh led directly from Lough Neagh to the head of Carling-ford Lough and the main seaport in north-east Ireland. The main factor that led to the construction of an effective channel of inland navigation between Newry and Lough Neagh was the desire and determination of Dublin interests to tap the recently discovered coal deposits between Dungannon and Coalisland in east Tyrone on the western shores of the lough. The link was forged between 1731 and 1742, the earliest artificial inland navigation of the canal era, and with the completion of the first ship canal, below the town, between 1759 and 1769, which afforded passage to large vessels, was of prime importance in contributing to the emergence of Newry as a major Irish seaport during the second half of the eighteenth century.

With the success of the Newry Canal, to the south, and the increasing exploitation of the coal deposits of east Tyrone, construction of a navigable cut linking Belfast with its lowland hinterland and the shores of Lough Neagh was not long delayed. However, though the head of navigation on the Lagan Canal reached Lisburn in 1763, the waterway was not driven through to Lough Neagh for another thirty years.

To the west the prospect of a substantial and sustained output from the new collieries of east Tyrone was sufficient to create a strong body of opinion in favour of linking the mining area with the shores of Lough Neagh and as early as 1732 construction work had begun on a short four-mile waterway down to the Blackwater. This Tyrone Navigation or, as it was frequently called, Coalisland Canal, was not completed until 1787, after the most appalling waste, procrastination, and inefficiency. As the output of coal from the mining area in subsequent years never reached the high level confidently anticipated in the mid-eighteenth century, this waterway enjoyed but moderate success throughout its entire existence.

At the beginning of the nineteenth century the idea of linking the lowland basins of Lough Neagh and Upper Lough Erne became popular with the more progressive and enlightened landed proprietors and merchants in Armagh, Monaghan and Fermanagh. The forty-two-mile-long Ulster Canal, constructed between 1825 and 1842 at a cost of over a quarter of a million pounds, made possible through communication by water from Belfast to Belleek and it was hoped that an extension of this line of navigation across County Leitrim to the headwaters of the Shannon would soon provide a third great line of water transit across Ireland, similar to the Grand and Royal Canals farther south. This final link in the east–west chain of navigations, the Ballinamore & Ballyconnell Canal was not completed until 1860 and proved a dismal failure. For this and other reasons the Ulster Canal, too, was unsuccessful, a severe financial embarrassment to its owners for the remainder of its existence.

Finally, between 1847 and 1859, the Irish Board of Public Works carried out a number of important improvements on both the Upper and Lower Bann, dredging and deepening the river course for the benefit of both drainage and inland navigation, constructing locks at salient breaks of slopes on the Lower Bann and erecting landing slips and small quays at many points around the shores of Lough Neagh and on the Blackwater, as well as on the channel of the Upper and Lower Bann itself.

The only waterway of significance to commercial navigation lying entirely outside the Lough Neagh basin was the short Strabane Canal, constructed between 1791 and 1796 from a prosperous market town, the principal centre of manufacturing and commerce on the Irish estates of the Abercorn family, to the navigable River Foyle at a point some ten miles upstream of the quays at Londonderry.

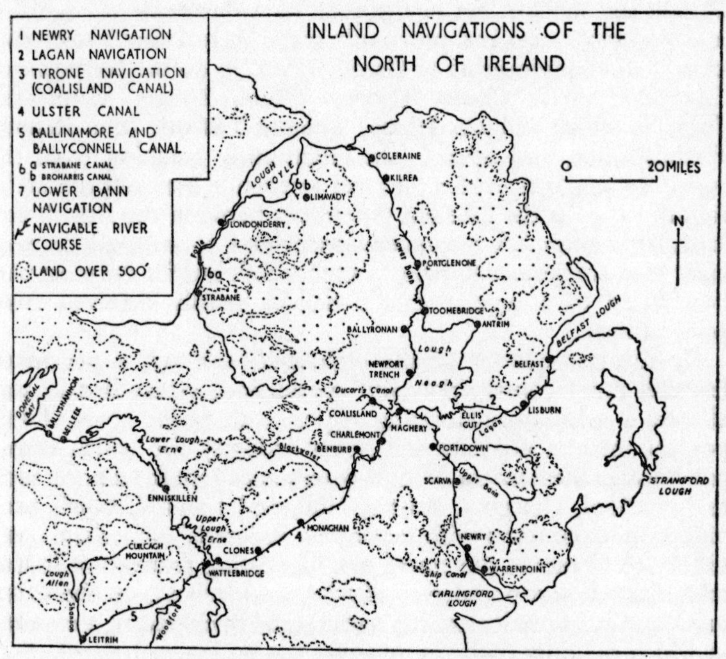

1. Inland Navigations of the North of Ireland

These, then, were the principal arteries of inland water communication in the north of Ireland. Their construction was complete by 1860, but many of them had already demonstrated the folly of expending huge sums of money on commercial waterways in an area where there was no regular, sustained traffic in minerals and heavy goods for which water transport was ideally suited. As elsewhere in the pre-railway age land transport could bear no comparison with water in regard to facility or cheapness, though in Ireland the absence of any great mining or manufacturing areas, and consequently of any major regions of high population density, greatly reduced the benefits generally resulting from large-scale inland water transport.

Moreover, though the earliest inland canal constructed in the British Isles lay in Ulster, it was built to the design and under the direct supervision of an Englishman. Indeed, in matters of canal engineering generally, as in so many other instances, Ireland suffered by being considered fundamentally similar to Great Britain and most of her navigations were the work of engineers and surveyors who had previously made their names by the success of their engineering in Great Britain, or on the continent of Europe—Richard Castle, Thomas Steers, Robert Whitworth, John Smeaton, William Jessop, Thomas Telford, John Rennie, Francis Giles, and others. The engineering resulting from the visits of these men was for the most part sound but, unfortunately, failure to grasp the inherent differences between the British and Irish economies left the several waterways to struggle on year after year, sometimes showing a slight profit but for the most part committed to a regular annual deficit, usually defrayed from public funds.

By the last quarter of the eighteenth century, a period of feverish canal construction and speculation, Great Britain was already a great manufacturing region with major areas of important mineral deposits and heavy industry for which canals provided an ideal means of inter-communication: from ports such as Liverpool, Glasgow, Bristol, Hull, and London, canals provided direct contacts with expanding industrial hinterlands and did much to mould the emergence of existing patterns of industrial location and population distribution. Ireland, on the other hand remained predominantly agricultural in character, a country in which the only traffic in minerals was an internal distribution from points of import: there were few raw materials, beyond the produce of the land, for which inland water transport was ideally suited, and the peasant population was quite unable to support regular passenger services.

Successful elsewhere, canals were constructed in Ireland simply because it was thought that such developments would create and stimulate traffic, as they had done in Great Britain, and in more than one instance merely to provide employment for a restless and destitute peasantry. Numerous lakes and navigable rivers often facilitated canal engineering, but this was insufficient to outweigh an overall lack of justification in the cultural environment. The majority of the lines of navigation were never able to command a traffic volume of anything like the dimensions required for commercial stability and with the rapid growth of the railways in the mid-nineteenth century their difficulties were further accentuated. (See diagram on page 115.)

For well over a hundred years canals such as the Newry and Lagan enjoyed moderate success, carrying agricultural produce and local manufactures to the ports for export or further processing and distributing fuel, grain, and raw materials over wide areas round the shores of Lough Neagh or along the river valleys radiating from it. However, when considered in direct relation to the total tonnages passing through the ports of Belfast and Newry, especially after 1850, canal traffic is seen to account for less than 10 per cent of the total. Only the Lagan and Newry were commercially successful and even with the latter prosperity was due largely to the success of the ship canal, especially after the extension to Upper Fathom, the erection of the great Victoria Lock and the partial clearance of the Carlingford bar in the mid-nineteenth century. The others for a time enjoyed moderate success but the expenditure of large sums of money was never rewarded by the wide range of benefits invariably promised in prospectuses or preambles to Acts of incorporation. In the case of the Tyrone Navigation, the failure of the collieries to provide a regular downstream coal traffic robbed that short waterway of much of its meaning. The Ulster Canal, laid out by Thomas Telford in the fashion of the period as an unusually direct line of water communication between the Blackwater and Upper Lough Erne, was soon in direct competition with the Ulster Railway extension from Armagh to Monaghan and Clones and this, coupled with foolish cheeseparing in construction and extreme difficulty in obtaining an adequate water supply, led to its rapid eclipse as a significant or self-supporting waterway. The Foyle Navigation (Strabane Canal), largely the brainchild of James Hamilton, Marquis of Abercorn, contributed to the prosperity of Strabane during the first half of the nineteenth century but its short hauls were particularly vulnerable to parallel and competing lines of road and rail communication and its commercial importance steadily declined as the century progressed. Lastly, the improvements effected by the Board of Public Works on the Blackwater, Upper and Lower Bann and Lough Neagh in the mid-nineteenth century came much too late to fulfil the glowing anticipations of the Commissioners in 1845. While Lough Neagh had for long been of great importance in linking east Tyrone, south Derry and north Armagh with the Lagan valley, providing opportunities for movement of population and economic contact, the rapid growth of lines of rail communication throughout the Lough Neagh basin soon resulted in the orientation of the entire area towards the expanding port and manufacturing centre of Belfast.

Apart from the ship canal at Newry the inland waterways of the north of Ireland have long since ceased to be commercially significant in the network of internal communication on which modern trade and industry depend. Their place has been taken, firstly, by the railways and, latterly, by road transport. The increased tempo of nineteenth-century commerce and industrialization found little satisfaction in the facilities for conveyance and travel provided by the internal waterways constructed during the previous century. As the railway network over the greater part of the province was later to disintegrate in the face of road competition, so these canals were unable to fulfil the transport requirements of the age of industry and after a brief half-century of moderate success entered a period of slow but uninterrupted decline. Those constructed in the nineteenth century had even less chance of success and traffic handled soon declined to negligible proportions.

Mainly the product of the eighteenth century and at first providing revolutionary services for the leisurely movement of a large variety of agricultural produce, raw materials and imperishable foodstuffs in an economic system hitherto dependent on primitive road transport, their eclipse was complete by the early years of this century. Since then barge traffic has steadily declined and the continued operation of the waterways of Northern Ireland reflected more the economic axiom of momentum than any grounds for survival in the severe competition of the modern world.

The collapse of the canal and rail systems in Northern Ireland has proved that the local economy, with no major internal traffic lanes, a dispersed rural population largely dependent on agriculture and but one great centre of manufacturing industry and international commerce is incapable of supporting systems of inland transport for which regular and sustained traffic between fixed terminals is an essential prerequisite.

CHAPTER II

The Newry Navigation

To the south of Lough Neagh the valley of the Upper Bann and the lowland corridor between the uplands of Down and Armagh lead directly to the head of Carlingford Lough and the important port of Newry. As early as the 1640s, during the Cromwellian campaign in Ireland, the area was surveyed and an order given by Colonel Monk for the cutting of a navigable trench linking Portadown with Newry. Nothing came of this project, however, and it was not until the beginning of the eighteenth century that Francis Nevil,

> Collector of Her Majesty's Revenue in Ireland . . . survey'd and level'd the Glan Bog lying between the Counties of Down and Armagh, as also the course of the Upper Band and Newry River so as to shew the whole extent between Logh Neagh and Newry . . . the same being taken with a designe of drawing a Canal or making a Passage for Boats from the said Logh to the Sea . . . being undertaken at the request of several Members of the Honble. House of Commons . . . 1703.[1]

Nevil estimated that a waterway navigable by lighters 'of twenty tons burthen' could be constructed for £20,000, but though a Committee of the Irish House of Commons was appointed in September 1703, to draft a bill for the purpose, nothing further appears to have come of this scheme.[2]

While no cut was made as a direct result of Nevil's survey, the discovery and exploitation of coal deposits in east Tyrone, on the western shores of Lough Neagh, in the years immediately following, soon led to pressure for a line of inland navigation by Dublin interests who wanted direct access by water to Lough Neagh and the valuable deposits of coal outcropping between Dungannon and Coalisland. The failure of an Act of 1715[3] 'to encourage the draining and improving of the boggs and unprofitable low grounds, and for easing and despatching the inland carriage and conveyance of goods from one part to another within this kingdom' and the general inability of local interests to provide the necessary capital or technical skill required in undertakings of this kind prompted the Dublin Parliament in 1729 to establish[4] the 'Commissioners of Inland Navigation for Ireland', a central body consisting of the Lord Lieutenant, the Lord Chancellor, four Archbishops, the Speaker of the House of Commons and eighty other responsible persons, to have complete responsibility for inland navigation in Ireland. The same Act levied duties on a wide range of luxury goods, from carriages, dice, and other gaming devices to silverware and wrought metal plate, to provide the new Commissioners with funds for their task.

By now public interest in the coal workings in east Tyrone had been thoroughly aroused and it was confidently predicted that with the establishment of water communication between the mining area and Newry, one of the chief seaports of the day, Ireland, or more correctly, Dublin, could well become independent of costly and unreliable cross-channel imports. Official sanction was given to the scheme, but work did not begin until 1731.

Some three years earlier Richard Castle had come to Ireland. He was probably a Huguenot refugee who had moved to Hesse-Cassel in Germany early in the century because of religious persecution in France, and had then travelled widely in Germany, the Low Countries, France, and England, studying navigation works. Upon arrival in Ireland he had taken a post in Edward Lovett Pearce's office in Dublin as an architectural assistant, and probably prepared his manuscript account of continental canal engineering[5] when Pearce, an architect of great flair and ability, was appointed Surveyor General in 1730 and thereby became directly responsible for implementing many of the suggestions and proposals for inland navigation of the Act of 1729.[6]

Castle had probably been asked to Ireland by Sir Gustavus Hume, who wanted a country seat built alongside Lower Lough Erne in Co. Fermanagh, and who had heard of Castle's ability as both an architect and engineer. As it happened, Sir Gustavus was one of the 'eighty responsible persons' appointed under the 1729 Act, and while Castle is primarily remembered as an architect, he soon became involved, with Pearce, in navigation works. At first he was entirely under Pearce's supervision, but when the latter died in 1733, he took complete control of the ambitious Newry project.[7] Indeed, it is probable that Castle, who was a capable but much less brilliant architect than Pearce, knew more about canal

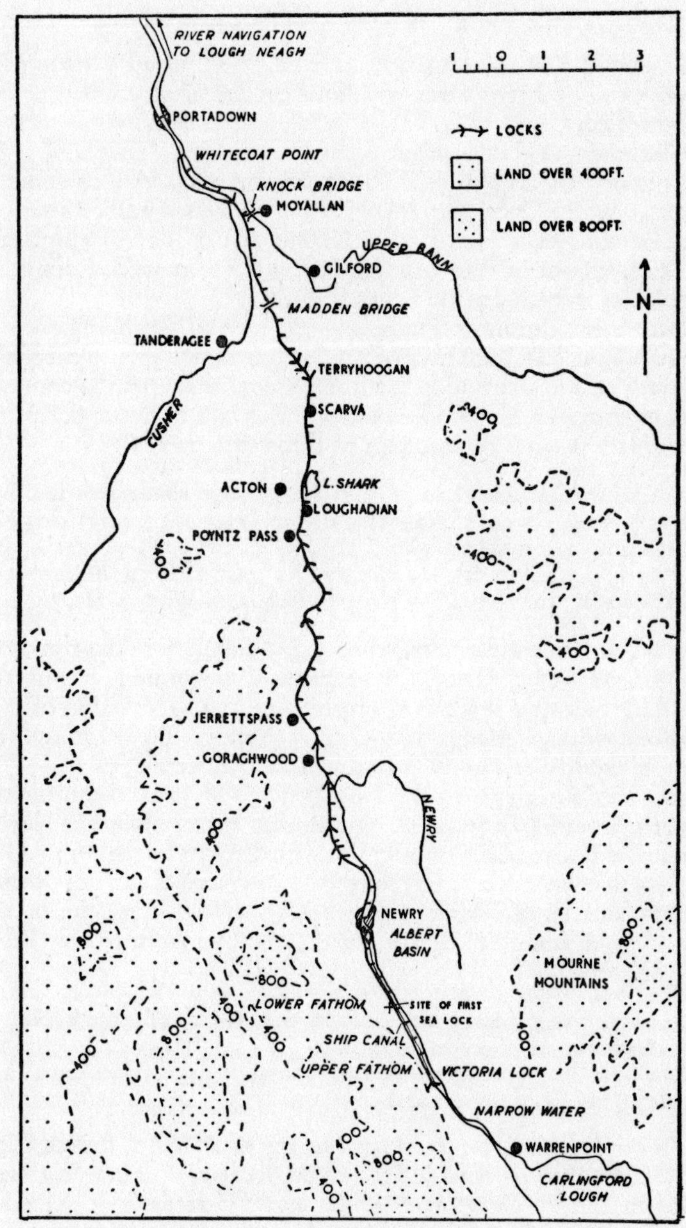

2. The Newry Navigation

engineering than his superior, and that he exerted a strong influence on whatever work was done on the Newry Canal before Pearce died.

Castle was in charge of the work for the three seasons of 1734, 1735 and 1736, and probably built the first stone lock chamber in Ireland, but in December 1736[8] he was dismissed and returned to Dublin with his Lisburn wife, Jeanne Trufer, also of Huguenot descent, whom he had married in 1733. He remained active in Dublin as an architect till his death in 1751.

Before his dismissal Thomas Steers[9], an English engineer of some repute, had been invited, in the spring of 1736, to carry out a survey of the area and in the following year accepted a commission to complete the canal, entering into a contract to superintend construction work over a span of three summers.[10]

> . . . by the last letters from England we have an account that Mr. Steers, an engineer of considerable note, is set out for this kingdom to finish the canal at Newry, which it is reckoned will be a very great advantage to this kingdom with regard to furnishing this city with coals, which we have hitherto had from Whitehaven.[11]

Steers had evidently envisaged no great difficulty in the construction of the Newry Canal, for he planned to combine it with the building of a new dock at Liverpool and the supervision of new harbour works at Ballycastle, County Antrim. However, unforeseen difficulties retarded progress and between 1737 and 1741 Steers was forced to spend long periods in Ireland during the summer months, supervising engineering works along the Newry cut and at Ballycastle harbour.[12] The final stages in the completion of the canal were delayed by legal disputes[13] and it was not opened for traffic until 28 March 1742, when the first colliers, the *Boulter* and *Cope* of Lough Neagh, arrived at Dublin from Newry laden with Tyrone coal.

> . . . on Sunday last (28th March) the 'Cope' of Lough Neagh, William Semple Commander, came into this Harbour laden with Coals, and being the first vessel that has come through the Canal, had a flag at her topmost head and fired guns as she came up the Channel.[14]

This navigation, the first major inland canal in the British Isles, was indeed a considerable feat of engineering.[15] Some eighteen miles in length, about 45 ft wide, and 5–6 ft deep, it extended from a lock to the south of Newry 'contrived to receive and shut out the tide as occasion may require' to its point of junction with the Upper Bann and Cusher rivers, commonly called Point of Whitecoat, about one mile south of Portadown, whence river navigation was possible northwards into Lough Neagh and beyond. The initial cut was made through Damolly, Poyntzpass, by Lough Shark, Scarva, Madden Bridge, and Knock Bridge, to the Upper Bann: there were in all fourteen locks on the inland canal, each 44 ft long, 15 ft 6 in. wide and from 12 ft to 13 ft 6 in. deep: each lock was faced with hard stone brought from Benburb in County Tyrone and the bottoms of the lock chambers were planked with deal timbers, two inches thick. There were nine locks south of the summit level, which was reached between Poyntzpass and Terryhoogan at 78 ft above sea level. Lough Shark, which lay just to the east of the canal on the head level, served as the main reservoir, though other streams also acted as feeders from time to time, the chief of these being the Camlough or Bessbrook river, the Pontzpass river, the Acton river, Reilly's feeder and the Tandragee feeder.

It was confidently anticipated that the new waterway would be of great benefit to the area through which it passed in improving drainage and in supplying a means by which coal from east Tyrone could be speedily and cheaply transported to Newry for regular coastwise shipment to Dublin:

> . . . the beds of coal with which the County of Tir-oen abounds, in the neighbourhood of a navigable lake, which has also a navigable connection with the sea by means of a canal, are of extraordinary consequence to the kingdom. The works are now in a fair way of answering the noble intention of the public in laying out above 50,000 l. sterling in order to open a free passage for supplies of fuel to a great part of the kingdom, but especially to the capital . . . this passage being a beautiful canal from the seaport of Newry to the navigable part of the Bann River which will inable the kingdom to receive coals at eight or nine shillings a tun at different places.[16]

Nor was it long before the navigation proved its worth during a period of acute social distress. During 1744 incessant rains had destroyed the crops over the greater part of the Lough Neagh basin and in the spring of the following year there was a threat of severe famine throughout much of mid-Ulster. It is recorded[17] that but for the newly opened Newry Canal, by which grain vessels of seventy tons burden could be brought from Newry into the heart of the province, some fifty thousand people would have perished from starvation. As it was, corn to the value of £150,000 was imported along the waterway above Newry at this time and the threat of famine removed. The linen manufacture, too, benefited greatly 'upon account of the convenient water carriage for kelp and all other weighty materials necessary for that valuable branch of trade . . .' while general marketing and commerce over a wide area soon derived great advantage from the new artery of communication stretching northwards from an important Irish seaport.

As the construction of the Tyrone Navigation (Coalisland Canal) proved so protracted (1732–87), no coal was carried directly by water from the mining area to Newry until very much later. There was, however, an expanding export of coal from Lough Neagh ports, to which it was carted by laborious land carriage, to Newry via the Upper Bann and the Newry Canal. Coal was not the only commodity conveyed on the canal, however, nor did trade on it consist of a one-way traffic in export goods. From its opening, Newry sent supplies of flour and up to 60 tons of oatmeal weekly, as well as a great variety of general merchandise, to areas adjacent to the navigation. For example, between 1748 and 1767 some 1,750 vessels, varying from 25 to 50 tons burden, went up the canal, paying over £1,800 in tolls.[18]

However, the navigation soon began to show many defects of construction. Winter flood water breached the banks at places where the soil was at all friable or soggy; part of the summit level was so narrow that boats were unable to pass one another; two of the locks, 'built from a French plan . . . with pipes in the walls', often gave trouble when the pipes burst and the sides and splays of these locks, having been built in brick, were by 1750 already mouldering away. A third lock had been built of ordinary rough stone and was proving unsatisfactory, while two others were leaking so badly that it was not uncommon for them to run dry overnight. Moreover, Lough Shark was proving insufficient as a reservoir for the summit level in which scarcity of water was already causing much concern.[19]

Acheson Johnston, 'undertaker and manager' of the canal during the first years of its existence, outlined these deficiencies before a Parliamentary Committee in March 1750.[20] He suggested various remedies, including the tapping of the Cusher river to augment the water supply to the summit level and the diversion of a mill stream into Lough Shark to maintain its surface level which, because of the greatly improved field drainage resulting from the construction of the canal, had fallen markedly since 1730. He also proposed lay-bys on the summit level to obviate difficulties caused by the narrowness of the waterway in this section and the straightening of meanders on the Upper Bann, especially at Portadown Bridge, which allowed a minimum of clearance and was a serious hindrance to water-borne traffic in times of flood. The cost of these repairs and improvements was estimated by Johnston at over £2,000.

At this same Committee another witness stated that it was important not to lose faith in the new canal, emphasizing that its opening had given great stimulus to coal mining in east Tyrone. An 'English gentleman', John Fletcher, 'of great experience in coal mines', had at this time joined a partnership with the Bishop of Down and others to work collieries at Drumglass, south-west of Coalisland, and also at Stewartstown, and had sent over a number of experienced miners from Derbyshire to provide skilled labour. Fletcher was confident that cheap water communication to Dublin would justify his investment of capital and it was imperative to him that the condition of the Newry Canal should not be allowed to deteriorate.

As the century had progressed, however, it had become obvious that if Newry was to maintain a position of importance as a port and manufacturing centre, the condition of the tidal channel below the town would require attention. In its natural state it was quite unsuited to commercial navigation and unless drastic improvements were carried out it seemed highly probable that Newry would be unable to benefit from her favourable geographical situation, controlling entry into the heart of Ulster by the lowland corridor separating the uplands of west Down and east Armagh. It was particularly important that shipping facilities to and from the port should be improved as soon as possible in view of the link between Newry and Lough Neagh recently established by the inland canal. It was not to supply Newry with coal that the inland canal had been constructed, nor indeed to provide a channel by which merchandise could be distributed to an elongated hinterland from that port, but rather as an extensive and important link in the line of water communication between the remote, inland mining area and the capital, Dublin. Without ease of export from Newry by coastwise vessel, the inland canal would lose much of its *raison d'étre*. It was therefore considered more important to create an effective navigation channel below the tide lock that marked the beginning of the inland cut than to improve the Newry Canal itself.

When in late 1755 the merchants of Newry petitioned Parliament[21] for assistance in making the river navigable for about two miles below the first lock, the petition was favourably received and £4,000 was granted to Robert Needham, Member of Parliament for the town and its environs and an important landed proprietor in the area. Work was begun in April 1759 under John Golborne of Chester, engineer of the Weaver Navigation, on a direct cut, 1½ miles long, through the marshes on the south side of the Newry river and by November 1759 this had been extended south-eastwards from the tidal lock by about a mile.[22] It was some 40 ft wide, 4 ft deep, and quite inadequate even for coastwise vessels.

It would appear that shortly afterwards there was public disagreement between Golborne and the Newry merchants of the day over the new channel below the town, which resulted in Golborne's return to England a little over a year after beginning work on the canal extension. It is not improbable that Thomas Omer, of Dutch descent, who had been appointed 'engineer' of the inland canal in 1754, had little time for his fellow engineer and was instrumental in having the entire Golborne scheme abandoned. At any rate, soon after Golborne's departure an entirely new ship canal was cut through the salt marshes and mud flats of the river estuary, again on the Armagh side but slightly farther south, under the direction of Thomas Omer and Christopher Myers. This extension, constructed under the auspices of several prominent local merchants and landed proprietors was some 60 ft wide, 12 ft deep, and had a large lock at its seaward end at Lower Fathom, 130 ft long and 22 ft wide, capable of handling vessels up to 120 tons.[23] One mile and 60 chains in length it was finished in 1769, at a cost of £23,000 and a new dock and piers were built at Warrenpoint, where ships of over 120 tons still had to tranship their cargoes. The old navigation by the river was still kept open for boats and small craft and a link between it and the new canal was established by a small lock at Newry.[24]

The growth of Newry as a manufacturing town and major port during the second half of the eighteenth century was greatly fostered by this seaward extension of the artificial navigation channel and linen, in particular, became an important manufacture and export, due in no small measure to the prohibition placed by the British Government on the export of Irish woollen goods. The export trade included 'a great variety of Irish linen, iron ware, leather shoes and boots, soap and candles, tea and refined sugar, beer and foreign spirits'.[25] Indeed, Newry's trade during the latter half of the eighteenth century rivalled that of Belfast and in 1777 the port ranked as the fourth trading town in Ireland, 'supplying most of the province of Ulster with foreign merchandise and having a great export of linen cloth, beer and butter'.[26] A direct passenger service with America was also established at this time, probably from Warrenpoint, while trade developed with ports as varied as Alicante, New York, Göteborg, London, Rotterdam and Liverpool.

To this general prosperity the Newry Canal had made no small contribution, acting as an artery of internal distribution for imports and an important flow line for goods being assembled at the port for export. However, in 1787, following expressions of dissatisfaction at the condition of the navigation and at inefficiency and apathy in operation and administration, the 'Corporation' for inland navigation, into which the Commissioners had been formed in the 1750s, was dissolved and the navigation vested in local interests.[27] This local control operated rather loosely through a central body or corporation of merchants and landed gentry in the Newry area for whom the commercial prosperity of the port was very much a practical concern. Members were obliged to retire in rotation but because of a general lack of competition many served continuously from 1787 to 1801. Day-to-day working of the canal, upkeep and maintenance during these years were in the hands of a canal superintendent, normally responsible to the local corporation.

Towards the close of the eighteenth century the continued lack of success of many of Ireland's canals and their inability to show any return for the very large sums of money expended during their construction prompted the Irish Parliament to adopt a general policy of decentralization, reflected in the transfer of control of many of the waterways to local trustees and commissioners. Unfortunately, these local interests had neither the experience nor the ability to reconstruct or overhaul canals which were already far from satisfactory and the position worsened as the century drew to a close.

In the case of the inland section of the Newry Canal the condition of the watercourse had by 1787 so deteriorated as to cause deep local concern, but despite ambitious schemes to extend the ship canal some four miles farther downstream and to build a link channel between the mouths of the Upper Bann and Blackwater, near the southern shore of Lough Neagh, no further improvements were carried out for many years. A Committee of the Irish Parliament was told in June 1800, however, that the raising of the banks of the ship canal to give an extra foot of depth was to be started later in the year under the direction of Richard Owen, engineer for the Lagan Navigation extension to Lough Neagh, whose services had been secured in November 1799.[28]

By the last years of the eighteenth century other connecting waterways had been opened. The Tyrone Navigation (Coalisland Canal) had been at last finished in 1787, by which time, however, in which its upper extension, Ducart's Canal, opened ten years earlier, had ceased to be used. The Lagan Navigation reached Aghalee in March 1792 and was driven through to Lough Neagh at Ellis' Gut in December 1793. The main downstream traffic on the Newry Canal during the final years of the eighteenth century comprised coal from Lough Neagh and east Tyrone, linen goods from the early mills of the Upper Bann and Lagan valleys, quarried material from mid and south Down, mainly slates, road metal and granite, grass seed and general farm produce, including meat, pork, dairy produce and roots. Cargoes moving inland from Newry to the small market towns on or near the waterway, Portadown, the Lough Neagh ports and the Lagan and Tyrone Navigations, included grain, miscellaneous foodstuffs, tools and machinery, foreign timber, tallow, chemicals and dried fish. By the end of the century, trade on the inland canal was yielding tolls of over £7,000 per annum and there were a dozen or more lighters plying regularly between Newry and Portadown.[29]

As we have seen, control of the Newry Canal was transferred to local interests in the 1780s, and this policy of decentralization was extended by the Irish Parliament in the closing years of the century. However, the transfers produced no improvement, and with the abolition of the Irish Parliament in 1801, the reorganization of the administration of Irish affairs and their control from Westminster, a new body was set up, the Directors General of Inland Navigation, five Government commissioners appointed by the Lord Lieutenant, to whom was given control of the majority of Irish canals. Based in Dublin, they took complete responsibility for all those canals which had been built under the auspices of the former Irish Parliament, including the Newry and Tyrone Navigations in the north, but not the Lagan, which had passed to a private company in 1779. The creation of the Directors General, directly responsible to the British Government for the allocation and expenditure of money voted from time to time to improve Irish inland navigation, was an attempt to put the finances of Irish canals on a sound basis, once and for all.

Once they had taken over the Newry from the local commissioners in 1800, the new body called for a report on the condition of the inland canal from one of their engineers, Henry Walker. Receiving an altogether unfavourable account, and an estimate of over £23,500 for repairs and new works, disagreement arose between Walker and the Directors General with regard to the extent of the improvements required on the canal, both above and below the town, and the two parted company late in 1801,[30] the actual cause being alleged faulty workmanship at a lock at Poyntzpass. Walker was succeeded by John Brownrigg, under whose direction the inland canal was largely reconstructed during the next ten years, many of the improvements and repairs effected under him being visible today.

When Brownrigg assumed control the entire inland navigation was indeed in a ruinous condition: the harbour at the seaward end, at Fathom, was suffering from severe silting, a wooden trough carrying the Cusher feeder was in a rotten condition throughout, and many canal banks on the inland section were liable to collapse:

> all the nine locks from Newry to Poyntzpass are little better than ruined makeshifts, pieced and patched from time to time these sixty years past; they were never right good, having been originally finished with brick and, as they failed, replaced with ill-cut or punched, ill-jointed mountain stone and filled up with bad rubble-work behind, except in the lower parts, under water, where some of them are of rubble stone of as good work as any in Ireland. The floors were of deal planks and the breasts and sills of oak. The gates and sluices are bad and on an ill construction, most of them now crazy and shaking, without copeing or upper heelstones to retain the iron land ties of the gates. The iron work of a most wretched kind is generally naked above ground instead of being bedded under the greatest stones that could be had.[31]

Acting on Brownrigg's report, which concluded that 'the repair of the canal of Newry and placing it in a respectable condition from end to end will prove little less expensive than making a new one of the same length', the Directors General embarked on an extensive programme of reconstruction and improvement, which included enlarging the locks, and between 1801 and 1811 over £38,000 was expended on both the ship canal and inland navigation, of which over £23,000 was a public grant. The improvements carried out included the widening and deepening of the summit level, the rebuilding of four locks and three bridges, the replacement of a double lock at the northern end of the summit level by a larger, single lock and the thorough overhaul of the lock at Lower Fathom, which formed the seaward end of the navigation.[32]

Not surprisingly the early years of the nineteenth century saw a temporary decline in traffic and receipts on the section above Newry, due both to the temporary closure of sections of the canal to permit repair work and also, perhaps, to a distrust of the change of control from local commissioners to Directors General. At any rate, in 1803 the number of lighters in use had diminished greatly and tollage receipts for that year amounted to only £2,500.[33] By 1811, however, the canal had been placed in a fairly complete state of repair and determined efforts were made to attract traffic to the newly improved waterways, which would also assist the agriculture of the area through which the canal passed. Thus certain products such as limestone, lime, sand and manures, were allowed transit on the inland waterway free of tolls. However, either these concessions were misunderstood or ignored, for it is recorded that farmers still carted their lime on the road which ran close to the canal and, in some places, almost parallel with it. By this time, too, the inland canal had come to act as a channel by which coal, imported at Newry, could be distributed to the bleach greens of the interior and also for domestic use in Down, Armagh and farther afield, a reflection of the meagre output of the native coal workings in east Tyrone at the beginning of the nineteenth century.

Passenger services were also now introduced. The *Newry Telegraph* of 17 February 1813 tells us that a passenger service from Knock Bridge to Newry had been introduced 'by the respectable Quakers of Moyallen', which would enable people from a wide area 'to go to Newry, transact their business and return home the same day with ease, comfort and convenience'. Fares were 3s 4d return in the first-class cabin, and 2s 1d return second-class. On 17 April a packet-boat over the same length of canal was scheduled to take about four hours and the advertised single fares were 2s 11d (first-class) and 1s 3d (second-class). Cheap day returns were issued on Saturdays at 3s 4d and 2s 1d. This regular passenger service, the only one ever operated on an Ulster canal, continued for about thirty years, though the numbers of passengers carried must have been fairly small and no mention is made of it in the report of the Railway Commissioners in 1838. On the opening of the Ulster Railway to Portadown in 1842, the morning train from Belfast made a connection with the boat at Portadown on three days a week.

With the completion of the Lagan Navigation in 1794 and its further improvement during the next twenty years, a continuous link by inland navigation was developed between Belfast and Newry, the former just entering on a period of rapid expansion in population, commerce, and industry, the latter already an important port which had the advantage of being nearer the capital, Dublin, than her new northern rival. Nevertheless, the recovery of the Newry Navigation from the serious decline of the early years of the century was slow. Carelessness, apathy, the misappropriation of funds characteristic of public works in Ireland in the mid-eighteenth century, and the closure of the navigation for repairs, had done much to destroy the canal traffic that had been established immediately after the completion of the waterway. Once lost, it proved difficult to regain.

Although the canal works had been put in a thorough state of repair, largely at public expense, the problem of silting at the entrance to the ship canal continued, and had attracted Brownrigg's attention during the engineering works undertaken there in 1806. Indeed, since the 1760s, when the lock at Lower Fathom had been built, the depth of water there had decreased by some four feet, and it was with increasing difficulty that shipping could negotiate the navigation channel immediately below the lock.[34] The Directors General therefore asked Brownrigg and another eminent engineer of the day, John Killaly, to make a detailed investigation with a view to a major improvement in the tidal approaches to the artificial channel. After a survey carried out mainly by Killaly, it was recommended that a canal should be built along the seashore to a lock and basin in Rice's Bay, at an estimated cost of £80,000.

Before a decision was taken, however, several other engineers of note were consulted. Alexander Nimmo proposed to alleviate flooding in the low-lying areas around Lough Neagh by providing an alternative outlet to the Lower Bann for surplus water in the lough. In 1828 he suggested removing the locks on the Newry Canal and cutting the bed of the Upper Bann to a depth which would reverse its flow, lead off excess flood waters from Lough Neagh and direct them down the Newry Canal to the sea. In this way not only would flooding be lessened but the scarcity of water then being experienced on parts of the navigation would be remedied. John Rennie, in 1830, pointed out that the development of steamships would soon make the extension of the ship canal to Rice's Bay too small and therefore unable to earn its keep,

The Chamber of Commerce of Newry,

As well on behalf of the Noblemen and Gentlemen, the Proprietors of Land in and near the Town, and in this part of the Province of Ulster, as of the Merchants, Traders, and other Inhabitants of the Town and Port of NEWRY, beg leave to submit to your attention the following

STATEMENT.

THE Town of Newry possesses a communication by water with the Harbour, through the Canal, for two miles, to the Sea Lock at Fatham, and from thence, by the Newry River, to the deep water at Warrenpoint, distant about five miles from the Town.

The Newry Canal extends from the Town for about twenty miles, to Lough Neagh, into the heart of the Province of Ulster, and affords communication from the Sea, by water, to the highly populous districts of Down, Armagh, Antrim, Tyrone, and Derry; and has proved of the greatest advantage to the Public, by bringing up our Imports to this now great Mart, and by affording facility for the export of the produce of the before-mentioned districts, and of this neighbourhood in particular.

The navigation of the River, from the Sea Lock to the deep water at Warrenpoint, is difficult, tedious, and frequently dangerous; it cannot be attempted with safety unless with a leading wind, or in calm weather, by towing—and in either case only at High Water. The Narrows, at Narrow-Water, can only be passed under similar circumstances, and the Tide there is so rapid that many accidents occur to Vessels in passing.

The entrance to the Sea Lock is obstructed by Shoals, which have of late years encreased so rapidly, that the water for a great distance below the Lock is four feet shallower than it was sixty years ago; and now oppose such an obstacle as cannot be removed by any exertion within the power or means of the Board of Inland Navigation to apply; and unless counteracted by considerable and expensive works, promise soon to close up the entrance of the Navigation.

The injurious consequences to our Trade, resulting from this state of our Navigation, are, that ten, or even fourteen days are frequently lost between the arrival of a Vessel at Warrenpoint and her getting up to her place of discharge at the Customhouse-Quay, at Newry—and an equal delay is frequently, from the same cause, experienced in her departure for Sea.

To remedy these evils, which have been sensibly felt for many years, the Act of the 32d year of his late Majesty, cap. 10, was passed, which, after reciting "that the Newry Canal had proved highly beneficial, and that the extending the said Canal from Fatham further to the deep water of the Sea would be highly advantageous," empowers the then Commissioners for carrying on the Newry Navigation to deepen, widen, and extend it by making a navigable Canal from Fatham aforesaid to the Sea near Ryland River—and in order to provide for the expense of making and maintaining same, the Commissioners were, by that Act, empowered to levy Tolls, not exceeding 20d. per Ton on the burthen or tonnage of each Vessel, Boat or Raft navigating said Canal; which Tolls so provided to be encreased would amount to a net sum of £4,000 a-year and upwards; the former Tolls, which still continue to be paid, amounting to a sum of £2,400 a-year, and leaving a net surplus to the present Commissioners of Inland Navigation of Ireland of £1,400 per annum.

This extension of the Sea Level would not only ensure the safe and speedy passage of Vessels up and down the Navigation, beyond the influence of those impediments, but would enable Vessels to get to Sea with the very same wind that now prevents them getting down the River, and would allow Vessels of much larger burthen to come up into the Town than can now approach it.

The facts already stated would sufficiently prove the great advantages to be derived to the Trade of this great manufacturing and agricultural Province from the proposed extension; but two most important articles, Grain and Live Stock, are most of all affected by the present state of our Navigation. The export of these articles of our produce from Newry daily encreases, as well from the improving state of this part of the Country, as from this being the most convenient Port in the Province for the Trade to Liverpool, and the other Ports on that part of the Coast of England. Many cargoes of Grain have of necessity been re-landed and re-shipped at great inconvenience, labour and expence, and some have been almost totally lost in consequence of those delays; and cargoes of Cattle have suffered severely from the same cause, and some of both been totally lost, with Ships, Crews and Passengers—when, but for the before-mentioned obstructions and delays, the passage to the destined Ports might have been made by those unfortunate Ships, and actually was made by other Vessels, laden with Cattle and Grain, which were as low down as Warrenpoint at the time those which perished were so delayed up the River.

3. The beginning of a statement made in March 1824 by the Newry Chamber of Commerce to the Earl of Gosford

and instead recommended the clearing of the navigation channel below Lower Fathom down to Warrenpoint.

The final plan adopted was a compromise, in that while it adopted Rennie's proposal for clearing the natural channel below Lower Fathom it also included a 1½ miles downstream extension of the ship canal to a deep water basin called Doyle's Hole. By this time the entire navigation had passed under the control of the Newry Navigation Company, which had been formed in 1829 under the chairmanship of the Marquis of Downshire.[35] Consisting largely of Newry merchants, a condition of its formation was that £80,000 should be spent over the next seven years on the scheme of improvement already formulated by Government engineers.

The first half of the project was carried out between 1830 and 1842 under Sir John Rennie.[36] Heavy rock blasting was necessary at Narrow Water and the entire channel from Warrenpoint upstream to Squire's Point was deepened by three to four feet. This work cost some £48,000. The second half of the scheme, the extension of the ship canal, was equally protracted and lasted from 1842 to 1850. It involved the deepening and widening of the old ship canal, the construction of an entirely new extension 1½ miles long, 160 ft wide and 15 ft 6 in. deep, from the existing sea lock at Lower Fathom down to deep water at Doyle's Hole, and the erection of a lock, 220 ft long and 50 ft wide, at its seaward end at Upper Fathom. This section cost some £40,000 and was completed in April 1850, when the newly opened Victoria Lock and Albert Basin handled their first vessels, including three steamers of over 500 tons burden.[37]

This lengthy period of expansion and improvement from 1830 to 1850 saw the construction of most of the engineering works which remain in use today. It bears witness to the determination of Newry interests to keep their port in vital contact with the open sea by creating an adequate navigation channel below the town and by providing navigational and dock-side facilities for the larger steam vessels already carrying the bulk of Newry's import and export trade. Before considering the extent to which these improvements were successful, however, we should look at the position of the inland section of the navigation in the economic life of the province after the major repairs and improvements of 1801 to 1811 had been completed.

The Directors General of Inland Navigation, under whose auspices these had been undertaken, lived mostly at some distance from the port and consequently had not the local contacts, or interest, necessary to maintain the works in a satisfactory condition for traffic. Whereas under the local commissioners who had managed the canal from 1787 to 1800, tollage receipts from a waterway in poor condition had been of the order of £6,000 to £7,000 per annum and ten to fifteen lighters had been in constant use on the waterway, during the first quarter of the nineteenth century, despite the major improvements carried out under Brownrigg, traffic fell away markedly. For 1825 to 1830 inclusive, for example, tollage receipts from the inland section of the navigation were as follows:[38]

Year	Tollage Receipts		
	£	s.	d.
1825	2,746	6	7
1826	2,296	19	9½
1827	2,074	2	10½
1828	2,320	8	4½
1829	2,660	1	1
1830	2,309	1	7

These receipts represent an annual tonnage of between 50,000 and 60,000.

In 1829 control of the canal passed once again to local interests in the form of the Newry Navigation Company and at once traffic and receipts showed considerable improvement:[39]

Year	Tollage Receipts			Tonnage
	£	s.	d.	
1831	2,414	3	10	70,749
1832	3,029	11	8	88,434
1833	3,054	11	2	89,165
1834	3,404	0	8	99,383
1835	3,477	6	5	101,514
1836	3,520	18	2	102,770

Tolls paid by laden barges moving upstream from Newry to various points on the navigation, and beyond, were as follows:

Destination	Tolls	
	s.	d.
Scarva	30	0
Madden Bridge (Tandragee)	36	0
Portadown	39	0
Lurgan	39	0
Ballyronan	39	0

I. Newry Navigation: (*above*), bridge at Jerrettspass, designed by Brownrigg and dated 1808; (*below*) early nineteenth-century aqueduct between Tandragee and Scarva, carrying a feeder from the Cusher River to the summit level

II. Newry Navigation: (*above*) early nineteenth-century lock-house at Lock No. 7 (Goraghwood); (*below*) Lock No. 12, Terryhoogan, at the northern end of the summit

Rates of toll were of the order of 3s to 4s per lock, about 8d per ton, both outwards and inwards.

By 1840 Newry's trade had expanded considerably, especially in agricultural produce, to which manufactured goods were entirely subordinate. Butter had become the most important agricultural commodity in the commerce of the port and supplies for export were forwarded to the town from Down, Armagh, Tyrone, Fermanagh, Monaghan, Cavan, Leitrim, and even Longford. The newly completed Ulster Canal had to some extent lessened the trade, for although it made internal communication easier, some towns, such as Enniskillen and Monaghan, preferred to send their produce direct to the Lough Neagh basin and Lagan valley. Trade on the inland canal showed a continued increase, however, and by 1846 yielded £4,059 4s 1d in tollage receipts from a total traffic of almost 120,000 tons.[40] The chief articles carried downstream were linen cloth, butter, meat, coal, bricks and tiles, and general farm produce; upstream traffic included grain and flour, flax seed, tobacco, timber, iron goods, whiskey, foreign foodstuffs, hides, and oil.

By 1858, however, traffic and receipts showed an abrupt decline. Tolls had by then shrunk to £1,832 and in another ten years still further to a mere £662.[41] The reasons for this rapid decrease reflect a number of factors resulting from changes in the demographic and economic structure of the north of Ireland and from the advance of technology and its application to inland transport, as well as from more purely local causes.

The Newry Navigation Company was so heavily encumbered with financial obligations brought about by the major engineering projects of the years after it was formed in 1829 that in spite of an increase in traffic on the ship canal after the construction of the new sea lock, and further improvements on the canal and navigation channel in Carlingford Lough, no dividend was paid by the Company until 1876 and then only with permission of the Public Works Loan Commissioners, under certain rigid restrictions. The annual revenue from the inland navigation from 1865 onwards never exceeded £1,000 and repayment of loans and payment of a dividend during the remainder of the century were made possible only by the substantial income from the ship canal. Such a financial return did not encourage expenditure on maintenance or new works and replacements. Therefore the condition of the canal slowly deteriorated as the century progressed and this in turn made it less and less likely that freight traffic would be attracted to it.

c

The decline of the inland navigation was, however, mainly due to the growth of railways. Until 1850 Newry's contact with its hinterland had been maintained entirely by canal and road transport, but with the establishment of rail connections about the middle of the century, a new and vital link was forged between the port and the interior. In 1852 the line between Dublin and Belfast was completed by the construction of the final section between Portadown and Dundalk through Goraghwood. This ran parallel to the inland waterway along the greater part of its course, and when Newry was linked to the main line shortly afterwards by a short branch, much of the canal's traffic was transferred to the railway. Further, in 1846 construction had begun of a line from Newry to Armagh to link up with the terminus of the Ulster Railway. This was not completed until 1865, but in 1855 a branch was built to the newly-opened Albert Basin in Newry, whereby coal could be transferred to rail and the amount of labour involved in conveying coal to interior markets reduced. When the rail link between Newry and Armagh was finally completed, the seaport gained improved access to a populous and progressive farming and manufacturing area.

Thus, by the 1860s Newry was linked by rail to its immediate hinterland, and to the Lough Neagh basin and the rapidly expanding port and manufacturing centre of Belfast. While traffic handled by the port in 1831 had totalled some 103,560 tons, of which 70,749 tons were either forwarded or distributed by the inland canal,[42] by 1888, although the total had increased to 363,558 tons, traffic on the inland canal had diminished to 33,500 tons.[43]

A third factor contributing to the decline of the inland navigation was the disappointing output of the Tyrone collieries after 1840, which supplied little more than local demands absorbed. In each of the years 1831 to 1833 some 11,000 tons were exported from Coalisland by water,[44] the greater part to Newry, but by 1880 only five of the numerous pits that were operating during the period of maximum production were still open and the output of coal from these primitive, shallow workings was negligible in comparison with what it had been half a century previously.[45] The great expectations of earlier years were never realized and the downstream traffic in coal which had developed on the inland navigation as the output of the collieries increased during the 1830s and 1840s fell away just as rapidly. It was eventually replaced by substantial coal imports through Newry, but this brought no increase in inland canal traffic for the coal was distributed mainly by rail.

As the century progressed the relative importance of the port of Newry declined in face of the rapid growth of Belfast as a port and manufacturing centre and the subsequent extension of its hinterland south-westwards along the fertile Lagan valley into areas which had previously shown a clear orientation towards the south-east. During the early years of the nineteenth century Newry had maintained its position relative to Londonderry and Belfast, but with the introduction of steam navigation and larger vessels Newry was at a decided disadvantage, even after the new ship canal extension, lock, and basin had been finished in 1850. As early as 1841 the shipping entering Belfast had amounted to 387,930 tons, while Newry had only 106,080 tons and Londonderry 84,087 tons: by 1850 the figures were 573,563 tons for Belfast, 149,406 for Londonderry and for Newry 118,728.[46] Thus, while Newry's actual tonnage had increased, its relative position had markedly declined. The decline of the port of Newry, the great population decrease of the 1840s, the contraction of the hinterland due to the growth of Belfast and the failure of the Ulster Canal to provide a feeder navigation, tapping the Fermanagh lowlands and Enniskillen, meant that the mid-nineteenth century was a period of increasing difficulty and decline for the inland waterway. By the 1880s, indeed, there was very little traffic between the Newry and Lagan Canals through Lough Neagh, the greater part of the traffic on the former being imports from Newry to Tandragee, Scarva, Portadown, the Lough Neagh ports and Coalisland, to which there was an increasing import of coal. There was also lighter traffic in timber and heavy goods from Newry to Monaghan, Clones and Belturbet by the Ulster Canal, canal rates to these places from Newry comparing favourably with those on the parallel and competing railways.[47]

The condition of the Ulster Canal, linking the Blackwater with Upper Lough Erne was, however, by now very unsatisfactory. A loaded lighter took some three weeks to pass from Newry to Monaghan in fine weather, and about six in time of frost; furthermore, the lighters on the Newry Canal were too large in length, breadth and draught to negotiate its narrow locks and shallow channel. This meant that a costly and time-wasting break of bulk was required at Moy or Blackwatertown, either by lightening the cargo or by transhipment.

The Newry Canal was at this time handling a sizeable downstream traffic. This arrived at Portadown by sail and steam from the Ulster Canal, Tyrone Navigation and Lough Neagh ports and was transhipped into lighters there before moving down the canal towards Newry, usually for export but sometimes for further redistribution to the fertile agricultural lands of the Newry corridor. Tolls were of the order of 4d or 5d per ton but the more usual method of levying tollage was per lock, 2s being the normal rate on each of the thirteen locks between Newry and Portadown. Thus a full cargo of 60 to 65 tons paid some 26s on a single journey, either upstream or down. In 1888 canal traffic on the inland navigation totalled 33,500 tons paying tolls of £831, while maintenance costs and management fees stood at £334, leaving a net revenue for the Newry Navigation Company from this source of under £500.[48]

It is worth noting that Alexander Nimmo's proposals of over half a century earlier were revived at this time in view of the continued flooding occurring around the southern shores of Lough Neagh. In 1882 the chairman of Portadown Town Commissioners suggested that by making the Upper Bann some 21 ft deep, its flow could be reversed and one-ninth of the water entering Lough Neagh would thereby be diverted to a much more useful purpose. Further, by removing the locks on the inland navigation between Portadown and Newry and ensuring a supply of water to the artificial cut by diverting into it the flow of the river Bann, a ship canal could be established. A lockless canal from Goraghwood to Lough Neagh was envisaged, to take vessels of four or five hundred tons, which would greatly facilitate the import of coal to Portadown from Newry. A total annual import of approximately 145,000 tons was contemplated, besides a considerable export, and it was only the fact that the proposed improvements would involve extensive works under the Scarva to Banbridge branch of the Great Northern Railway and were therefore bound to be bitterly opposed, that had prevented a survey. The cost of this momentous engineering project was estimated at £200,000 which, it was anticipated, could be easily recouped from those between Scarva and Newry who would benefit from improved drainage, and also from tolls and dues on traffic.[49] Not surprisingly the scheme again proved too ambitious for local interests and was never undertaken.

In 1884, the Newry Navigation Act provided for the deepening of the upper reaches of Carlingford Lough and the Newry river for three miles, the channel of the river from the Victoria Lock to

Warrenpoint to be 120 ft wide, thus accommodating vessels of up to 5,000 tons. This work, which cost some £55,000 was carried out during the next ten years and proved to be the last scheme undertaken by the Newry Navigation Company.

4. The steam lighter *Ulster* at Newry about 1893. The craft is fitted with Henry Barcroft's large partially immersed propellers

In 1900 a bill was proposed by which the narrow-gauge Glogher Valley Railway from Maguiresbridge in Co. Fermanagh to Tynan in Co. Armagh would be extended through Armagh and Keady to Newry, but its supporters realized that competition from the Newry Navigation Company, which was by now quite independent of the town of Newry in its constitution, powers, authorities and management, might hinder the development of traffic. Further, there had been considerable friction between the town and the company for some time, principally over tollage rates and dues, and thus the relationship was not likely to be improved by the construction of a competitive narrow-gauge railway running parallel to the inland navigation over much of its course. Therefore the project for the rail connection was dropped and instead the Newry Urban Council and the railway promoters approached the company with a view to buying out their interests and embodying them in a public corporation on which the commercial interests of the town and port would be more adequately represented. After negotiations, the Newry Navigation Company was dissolved and replaced by the Newry Port and Harbour Trust, established in 1901. The new body, consisting of representatives of local merchants, ship-owners and ratepayers, was made responsible for 'the maintenance and improvement of the port and harbour of Newry, the Newry River and the Newry Canal' and thus the navigations were once more firmly in the hands of a public body on which local commercial interests were paramount. Though hampered by their financial position, much of it the legacy of past mismanagement, and heirs to a declining line of inland water transport and a sea communication inadequate to cope with modern shipping, the Newry Port and Harbour Trust took on substantial commitments for improvements mainly to the ship canal, basins and works on the tidal section.

As the century progressed the inland navigation proved a heavy liability. Constant repairs were required to keep it in navigable order and each year the annual reports of the trustees refer to the need for shoals and weeds to be removed from the navigation channel, and for attention to lock gates. However, the strenuous efforts made by the Trust to keep the canal in working order found no compensation in increased tonnage. By 1933, for example, traffic had diminished to 6,600 tons, by 1934 to 4,300 tons, by 1935 to 1,970 tons and by 1936 to less than 600 tons. Here are some figures for receipts:

Receipts and Expenditure on the Newry Inland Canal under the Port and Harbour Trust[50]

Year	Receipts from Tolls			Expenditure		
	£	s.	d.	£	s.	d.
1900	653	9	3	427	6	8
1905	844	0	0	474	0	0
1910	873	10	8	665	3	5
1915	646	12	5	601	8	5
1918	533	17	4	659	4	10
1922	803	6	1	1,125	14	0
1928	733	18	10	1,001	12	4
1932	406	14	3	1,199	16	2
1934	233	7	6	1,229	9	1
1939		17	6	711	16	4

The canal was derelict from 1939 and a warrant for abandonment as far as the town of Newry was issued on 7 May 1949, and through the town on 21 March 1956. The swing bridge on the Dublin Road which gave access to the inland navigation from the Albert Basin was replaced by a fixed bridge in 1958 and passage between the ship and inland canals then became impossible.

The Newry Port and Harbour Trust are still dependent on the Bessbrook river for water to service the ship canal. The river enters the inland canal approximately $1\frac{1}{2}$ miles above the first lock, so that a channel to carry the flow must be maintained by or on behalf of the Trust. The remaining 17 miles of the inland navigation is now of no value to the Trust. It functions now merely as a drainage channel for the lands on each side, and it is therefore probable that it will soon pass under the control of the Ministry of Agriculture in the Government of Northern Ireland. Though derelict, the inland section of the Newry Navigation remains of interest—to the Newry Port and Harbour Trust as, in part, an indispensable feeder for the ship canal; to local farmers and land-owners for drainage and 'rights of convenience'; to the Government of Northern Ireland in view of the uncertainty of future control and maintenance; to the County Councils of Armagh and Down because of their general concern with drainage and land improvement problems; and, not least, to the economic historian, for whom it reveals the aspirations and fluctuating fortunes of nearly two hundred and fifty years.

CHAPTER III

The Lagan Navigation

FROM as early as 1637 the commercial advantages to be gained by linking the town of Belfast with the shores of Lough Neagh by a navigable waterway had been in the minds of the more progressive merchants and landowners in east Ulster.[1] Such an undertaking would seem to invite the engineer, for the largest area of inland water in the British Isles is separated from the Lagan at Moira by only six miles of comparatively level ground and by linking the two, trade between the fertile lower Lagan valley and Belfast, on the one hand, and the Lough Neagh basin and beyond on the other, would be encouraged.

It was not, however, until the exploitation of coal deposits in east Tyrone and the construction of the Newry Canal between 1731 and 1742 that such a link between Belfast and Lough Neagh became immediately important. In 1741 a line was surveyed by Arthur Dobbs, His Majesty's Surveyor General, but it was not until 1753 that a petition in favour of a waterway was made to the Irish House of Commons by 'the sovereigns and burgesses of Belfast and Hillsborough, the inhabitants of Lisburn and the seneschals and grand juries of the manors of Killultagh and Moira'.[2] The weight of public opinion and commercial interests was now considerable and an Act was obtained on 24 October 1753[3] 'for making the River Lagan navigable and opening a passage by water between Lough Neagh and the town of Belfast in the county of Antrim'. Acheson Johnston, engineer in charge of the Newry and Tyrone Navigations at this time, had already carried out a survey, and given an estimate of £20,000 for making the river Lagan navigable to Spencer's Bridge, near Moira, and cutting a canal thence to Lough Neagh.[4] Such works as this were at this time in the hands of the Commissioners of Inland Navigation for Ireland, established in 1729, but as this body was as yet operating on a fairly restricted financial basis, provision was made in the Act to raise funds from the area through which the canal would pass, by levying an additional duty of one penny a gallon on ale and of fourpence a gallon on spirits, 'within that part of the district of Lisburn commonly known and distinguished by the Gaugers' walks of Belfast, Lisburn, Moyra and Hillsborough'[5] during the eleven years covered by the Act of 1753. Thus fortified, the commissioners saw fit to begin work on the scheme in 1756.

Within a year a navigable channel six miles in length extended upstream from Belfast. Four locks had been completed, two bridges and four lock-houses built, several miles of cuts made to shorten the river channel, most of the shoals removed, the river course dredged and the towpath constructed as far as Drumbridge.[6] The engineer in charge was Thomas Omer, who had been appointed superintendent engineer of the inland portion of the Newry navigation system in 1754 and after John Golborne's departure in 1760 became chief engineer for the first ship canal, below the town. With additional grants of £4,000 from the Irish Parliament in 1759 and again in 1761, work on the Lagan Canal proceeded apace.[7] Eventually, in September 1763, the navigation between Belfast and Lisburn was completed and opened to traffic amid scenes of great enthusiasm. The *Lord Hertford*, a lighter of 60 tons burden belonging to Thomas Greg, a prominent Belfast merchant, made the first voyage upon the new canal to Lisburn with a cargo of 45 tons of coal and timber:

> Mr and Mrs Greg had upon this occasion invited a numerous company of ladies and gentlemen to make the voyage and to dine on board. The day was indeed a happy one, the weather was fair, the prospect diversified with bleach greens breaking in upon every reach of the river, together with the woodlawn and meadow and the happiness and jollity of the reapers in nearly every field cutting down their harvest, all diffused joy and pleasure. The party were met at Drumbridge by the principal gentlemen of the town of Lisburn who came on board and the whole company were entertained by Mr Greg and his family in the most elegant and polite manner with a cold collation and wines of all sorts. A band of music played the whole way to over one thousand persons who accompanied the lighter on the banks of the waterway as far as Lisburn where the inhabitants expressed their unfeigned satisfaction at the completion of this great and truly useful work up to their town and at the near prospect of its being rendered much more advantageous by having the passage by water opened to Lough Neagh.[8]

Between 1763 and 1765 the river was made navigable up to

what later became the site of the Union Locks at Sprucefield, south-west of Lisburn, but here construction ceased. Between 1756 and 1767 over £40,000 had been spent out of the public navigation fund alone and Omer was experiencing considerable financial difficulty in continuing engineering works above and below Lisburn. In addition, it had become apparent that there were serious drawbacks in using the course of the river for sections of the navigation, especially below Lisburn, in that the Lagan was subject to severe winter floods, while the powerful vested interests of the linen bleachers were sufficient to prevent the navigation commissioners from having full control of the river's flow.[9]

In 1768 Robert Whitworth, an assistant of the famous canal engineer James Brindley, was approached by a committee representing responsible opinion in the Hillsborough area and, on request, he surveyed the area and reported.[10]

The work already completed on the lower Lagan he regarded as basically unsatisfactory, mainly because of the unsuitability of the river for commercial navigation. Whitworth's proposals were as follows:

(a) to abandon the river course altogether, because of the fall of 80 ft between Lisburn and Belfast, and to construct an entirely new canal, quite separate from the river, either via Malone and Dunmurry or by 'Strandmills' and Drumbridge, at an estimated cost of some £22,000;

(b) to extend the navigation beyond Sprucefield either by the construction of a canal on the north or Antrim side of the river, the course which he favoured, or by the extension of the existing waterway as a canal on the south or Down side, to cross the Lagan by an aqueduct at Spencer's Bridge near Moira.

These suggestions were well received and serious attempts were made to raise funds to complete the canal, mainly by increasing local responsibility and making it easier to borrow money. By Acts of 1772 and 1774[11] local commissioners, consisting mainly of landowners in the district through which the canal passed, but including a number of Belfast merchants, were established and empowered to borrow up to £10,000, offering as security the tollage receipts and the funds to be collected under the spirit tax. These were replaced in 1779[12] by the 'Company of Undertakers of the Lagan Navigation', in which the Marquis of Donegall held a

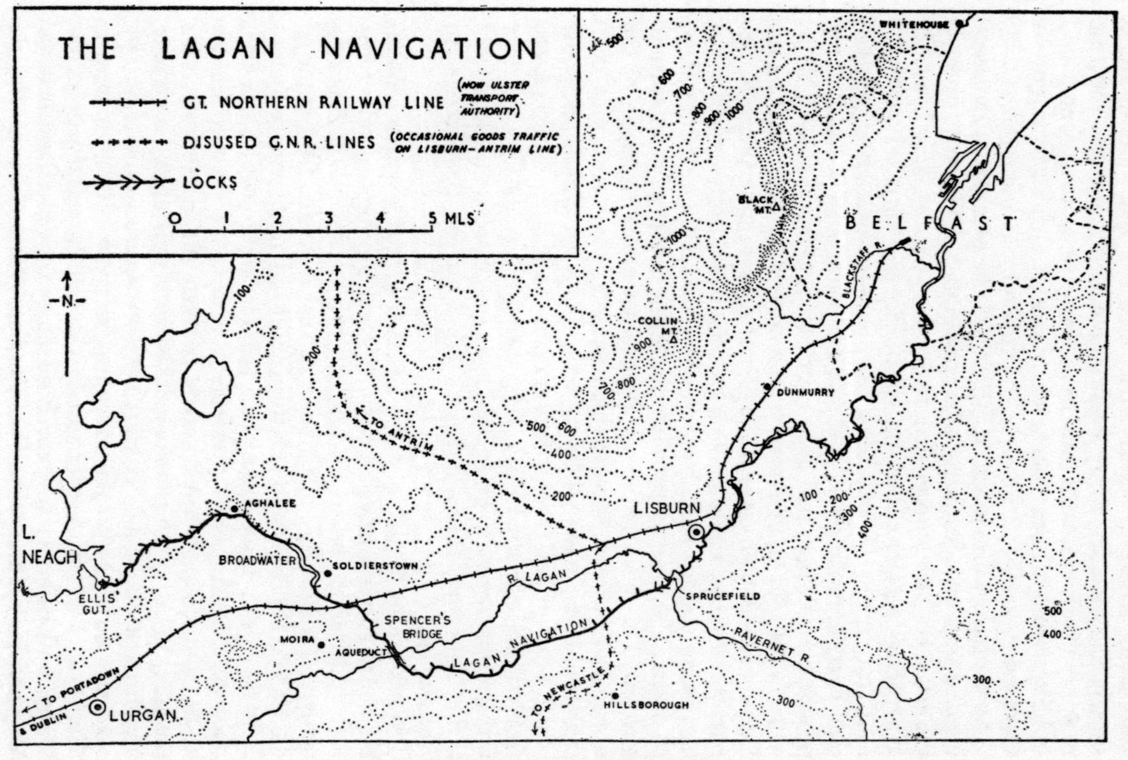

5. Map of the Lagan Navigation

controlling interest. Significantly, whereas the local commissioners had included many small landowners and merchants, the new company consisted almost entirely of landed gentry and nobility, from whom substantial financial patronage might be expected.

Soon after its formation, in June 1782, the Marquis of Donegall entered into an agreement with an English engineer,[13] Richard Owen, who had worked for some ten years on the Leeds & Liverpool Canal, to come over to superintend the canal works for four years at a salary of £200 a year, but although work began in that year, completion of the navigation proved a much more protracted task than had been anticipated. Owen favoured Whitworth's plan for an extension of the navigation by a canal on the Down side of the river, for in spite of considerable financial support from the Marquis of Donegall, the new Company had as yet insufficient money to embark on the more costly northern cut, or to replace the river improvements below Lisburn by an entirely new canal in this area. It was on the headward extension of the navigation beyond Sprucefield, therefore, that work was concentrated in the years following 1782. From the Union Locks, by which boats passed from the river on to the lengthy summit level, the final line ran almost due west through County Down before turning north-west to cross the Lagan near Spencer's Bridge over a fine aqueduct, through Friar's Glen, or the Broadwater, to Aghalee and thence down to Lough Neagh.

Work on the headward extension of the canal began in December 1782. The construction of the Spencer's Bridge aqueduct was first undertaken and in December 1782 we find Owen inviting tenders, the aqueduct to be built 'of that most excellent stone of the Earl of Hillsborough at Kilwarlin',[14] quarried some two miles from the site. It took about three years to build, cost £3,000 and, once completed, Owen pressed on with work on the summit level, both downstream towards Lisburn and north-westwards towards Aghalee, which was reached in March 1792. From here merchandise arriving by canal was for the time being conveyed by road to a little bay on the south-eastern shore of Lough Neagh, Ellis' Gut,[15] whence barges sailed to the south-west corner of the Lough and entered the river Blackwater to serve the towns of Moy and Blackwatertown. Coalisland was reached by the Tyrone Navigation, completed in 1787, and coal from here was carried downstream to the glass works at the Long Bridge in Belfast. At this time, too, the Lagan Navigation was used to convey large bottles and similar glass containers from this factory to the sulphuric acid manufactory on Vitriol Island, Lisburn, established by Thomas Greg and Waddell Cunningham in 1764.[16] It was not until December 1793, however, that the canal was driven through to the lough shore at Ellis' Gut and on New Year's Day 1794, the new headward extension was formally opened by the principal shareholder, the Marquis of Donegall, who made the trip from Richard Owen's house, near Moira.[17]

The final section of the navigation had cost some £62,000 and had been constructed almost entirely at the Marquis's expense. As well as the Spencer's Bridge aqueduct, it included the eleven miles of summit level between Sprucefield and Aghalee, the ten large locks, each 70 ft by 16 ft, that took the navigation down to the lake over a distance of $3\frac{1}{4}$ miles, the impressive flight of locks at Sprucefield that lifted it some 26 ft from the river to the summit within less than a hundred yards, several bridges, and a number of small basins and stores, including Newport, Kesh, Nocher's Quay, Harvey's Quay, Moira, Soldierstown, and Aghalee.

Lough Neagh had at last been linked to Belfast by a major waterway and it was confidently anticipated that commerical interests, industry and agriculture throughout the Lagan valley and much of the Lough Neagh basin would benefit directly. Unfortunately, however, the navigation proved a severe disappointment. That bane of early canal engineers, lack of water on the summit, proved a serious problem, as the Lagan could not be tapped indiscriminately because many of the early bleach greens of the valley depended on a steady and assured flow. At other times Belfast merchants, in particular, complained about the lower section of the navigation, where the artificial banks of the waterway were frequently carried away by the swollen waters of the Lagan in those stretches which followed its natural course. Due to the varying depths of water available, traffic was often delayed in the channel or at the locks.

Although the canal had been officially opened, it was by no means complete and work continued at the Union Locks until 1796. Already, however, enthusiasm had waned considerably and Owen found little support for any further work on the navigation. Indeed, the delay that had occurred in completing it and dissatisfaction at much of what had been done created some embitterment between Owen and his employers. It was hinted that he had 'protracted the business shamefully and for his own emolument' and that 'in Lord Donegall's precarious state of health and after the many delays and disappointments he has experienced, it will be no easy business to engage his Lordship further in an undertaking that has turned out so ill and contrary to his seduced expectations'.[18]

The nineteenth century opened most inauspiciously for the new waterway and in 1802 it was described as follows:

the good effects of the canal from Lough Neagh to Belfast are, in great measure, lost by the management of that part which runs from Lisburn to Belfast. The original defect seems to have been in the idea of making the river Lagan navigable, which, having a fall of so many feet and being a mountain river subject to many floods, is the most unfit that can be imagined for the purposes of navigation, being a great part of the year rendered so by the bursting of its banks, whose breaches cannot be repaired without draining a level and interrupting the passage of boats . . . had a canal been executed between the towns above mentioned without any connection with the river and with the same skill as the part latterly finished, the benefits would have been great, indeed, and the communications certain and speedy; but as matters are situated at present, more time is consumed in passage up the river Lagan, when there is a passage, than would be sufficient for the whole on a well executed and level cut.[19]

Whereas it had originally been hoped that the completion of the lough cut would link the colleries of east Tyrone with the Belfast district, in fact the Lagan Navigation soon became the chief means of distributing coal imported at Belfast to its immediate hinterland and to the lough shores beyond. Coal and timber were the chief goods moving upstream from Belfast at this time, and tiles, fireclay goods, and earthenware from the Coalisland area, sand, stones, and road metal the chief items being carried down towards the coast.[20]

In 1810 control of the Company of Undertakers passed from the Donegall family to a number of prominent Belfast businessmen and merchants who proposed to construct an entirely new canal from Belfast to the Union Locks at an estimated cost of between £40,000 and £60,000. Owen estimated a sum of £42,425[21] as the cost of separating canal and river below Lisburn, and on receiving an application from the controlling interests requesting a grant of half this amount, the Directors General of Inland Navigation[22] sent their own engineer, John Killaly, to inspect the works and report on the improvements proposed by Owen. The main difference between Owen's original proposals of 1810 and Killaly's

later ideas of 1812 was that the latter proposed a length of canal
from the mouth of the Blackstaff river to above the second lock, a
cut which would avoid most of the tidal section of the navigation.[23]

In 1812 the Rev. John Dubourdieu of Annahilt, near Hillsborough, commented incisively on the difficulties facing the navigation and on the new works proposed. Noting two main defects
in the navigation as it then existed, a liability to flooding and
breaching of the banks in the natural stretches below Lisburn and
an acute shortage of water on the head level during the summer
months, he continued:

> the plan for remedying the first error is to preserve as much of the
> navigation between Lisburn and Belfast as is independent of the
> river, making new cuts wherever they shall be required and only
> touching on the river in two places . . . and for the second, to turn
> the whole of the river Lagan, taking it up at Magheralin, into the
> head level.
> Both of these schemes, if executed, seem fully adequate to the
> proposed ends but unfortunately they clash with the interests of a
> body of men, the bleachers, which are so interwoven with the
> interests of the community that to them the utmost attention is due
> as they carry on the bleaching business to an immense amount along
> the banks of the Lagan, by whose waters their machinery is worked
> and which would be much injured by directing them into the head
> level, as well as all the other mills that lie on the same water, amount-
> ing to eighteen in number. Much injury, it is said, will also be sus-
> tained by them, even in making the new cuts from Lisburn to Bel-
> fast. . . . Besides the plan abovementioned of supplying the head
> level from the Lagan, Mr Whitworth proposed to take a stream
> which runs into that river at Costley's Bridge [the Ravarnet] by
> means of a cut into the head level, which could without great ex-
> pense be done, his estimate of the cutting and sluices necessary being
> in the year 1768 not more than 316 l. . . .
> Ideas have also been started of supplying the canal by pumping from
> the river with a steam engine, or even from Lough Neagh. Of its
> efficiency to perform so great a work I cannot form a judgement but
> Lough Neagh being nearly seventy feet below the level, one pump
> could not perform the operation. Upon the whole, it is a circum-
> stance much to be desired that some plan could be fixed on which,
> without much private injury, could make this navigation more cer-
> tain so that it should not in summer be interrupted by drought, nor
> in winter by floods.[24]

Most of the proposals advanced by Owen and Killaly were

Rise *of the* Locks *from* Belfast to the Head Level

		Feet	In.
N.º 1	Black staff River rises	4	6
„ 2	Morelands Meadow	5	0
„ 3	Shaws Bridge	8	0
„ 4	Norwoods	4	0
„ 5	Ballydrain	7	0
„ 6	Drumbridge	5	0
„ 7	Mc.Quistons	6	6
„ 8	Ballyskeagh	10	9
„ 9	Lambegg	6	0
„ 10	Bradleys	10	0
„ 11	Mc.Alice	9	0
„ 12	Lisburn	6	0
„ 13	Intended New Lock	5	1
„ 14	New	5	1
„ 15	New	5	1
„ 16	New	5	1
„ 17	New	5	1
„ 18	Old Lock the highest of the four connected Union Locks which raises the Vessels out of the Old into the new Navigation	6	8

Total 112 . 4

Whole rise from Belfast to the Summit Level.

6. The locks from Belfast to the summit of the Lagan Navigation, as stated by Richard Owen in 1813

III. Newry Navigation: *(above)* Merchants' Quay, Newry, in 1955, showing the grain stores and warehouses; *(below)* the wharf at Madden Bridge, looking southwards towards the summit

IV. Newry ship canal: (*above*) Victoria Lock at Upper Fathom, at the seaward end; (*below*) a collier moves down the upper section of the canal

V. Lagan Navigation: (*above*) sand barges at the Queen's Bridge, Belfast, about 1890; (*below*) tug and lighter on southern Lough Neagh, 1938

VI. Lagan Navigation: (*above*) below Lisburn in 1947; (*below*) an attentive audience at Drumbridge Lock, Upper Malone, Belfast, 1951

embodied in an Act obtained in 1814.[25] This included several proposals to divorce river and navigation channels between Lisburn and Belfast, most of which were never carried out, and also empowered the company to construct a reservoir in the Broadwater, near Aghalee, to ensure an adequate supply of water for the summit level. On the transfer of control in 1810 it had been suggested by Owen that the only method by which this problem could be solved would be to divert the flow of the Lagan from a bleach and corn mill near Magheralin into the canal and to purchase its water supply. This proposal aroused a violent storm of protest among the mill owners along the waterway and had to be abandoned. A reservoir was therefore authorized as an alternative, but was not built. Within a few years a large number of improvements were carried out, including the building of several sections of towpath, cleaning and deepening portions of the river channel, and repairing locks and weirs. As a result traffic greatly increased,[26] though water supply problems remained, and there was a horse-towing path for only part of the length of the navigation.

Barges and lighters could now operate throughout the year between Belfast and the Lough Neagh ports and beyond and traffic was encouraged by a 25 per cent reduction in tollage rates. Further, it was announced in January 1813[27] that to fill empty boats returning to Belfast no tolls would be charged on potatoes, hay, and straw moving downstream, thus encouraging farmers to avail themselves of cheap water transport for these bulky commodities. The improvements carried out during the decade beginning in 1810 resulted in a general acceleration of canal traffic and by 1820 the time taken for a loaded lighter to go from Belfast to Lisburn had been reduced to fourteen hours, Lough Neagh being reached in twenty-eight hours. These times compared most favourably with those usual at the turn of the century when a local anecdote held that when the canal was first opened a vessel made the round trip to the West Indies during the time taken by a lighter to pass from Belfast to Lough Neagh. Bradshaw's *Directory* of 1819 tells us that 'during the year ending 5th January last, 196 boats, of from 40 to 50 tons burthen, passed laden from Belfast to Lough Neagh; 63, principally laden with coals and lime, passed from Belfast to the summit level; 202 from Belfast to Lisburn, and 18 from Belfast short of Lisburn. And 63 lighters arrived, laden principally with grain, from the lake to Belfast; 28 from the summit level to Belfast.'

Despite the improvement, however, the navigation remained

D

unable to realize a net profit for its proprietors and with the refusal of the Exchequer Bill Loan Commissioners to advance money on the grounds that the canal was not sufficient security, the continuing scarcity of water on the summit level and the increasing animosity of the mill owners and linen bleachers of the valley,[28] all hopes of executing new cuts near Lisburn or of constructing a new canal course from a point on the first level, between the first and second lock, to the tidal reaches of the Lagan near the entry of the Blackstaff river, had to be abandoned.[29]

As the century progressed traffic increased considerably. Between 1831 and 1837, for example, 265,750 tons were carried on the waterway and tollage receipts for the same period totalled over £10,000, the usual tolls being 9½d per ton for the entire voyage of 25¾ miles.[30] The chief interior landing places were the bleach greens, mills and factories between Belfast and Lisburn, the lime-kilns and brickfields near Lisburn, the towns of Hillsborough and Moira, and the distilleries, breweries, and lime-kilns in their vicinity, the villages of Magheralin, Soldierstown, Aghalee, and Kilmore and the town, distilleries, and breweries of Lurgan. The principal goods carried upstream were coal, foreign timber, herrings, salt, groceries, iron, bleaching stuffs, spirits, barm, and bark and downstream, grain and flour—over 35,000 tons between 1830 and 1837—potatoes, sand, stones, firebrick, and tiles.[31]

This improvement was the result of the interaction of a number of factors: the substantial reduction in tolls, a further servicing of the waterway below Lisburn in the early 1830s which included dredging of the river bed, repair of the canal banks and a general strengthening of lock and flood-gates,[32] the general expansion of industry in the lower Lagan valley and the prospect of contact by water with the Erne lowlands and the Shannon basin which the Ulster Canal, then nearing completion, seemed to ensure. Lisburn was becoming increasingly important as a port of call on the inland waterway, a development reflected by the withdrawal in the 1830s of all dues from quays and landing slips there and in the erection of a large dry dock to facilitate the repair and servicing of the larger lighters now using the navigation. Lighters carrying 60 to 80 tons were now plying regularly up and down the waterway, many of them belonging to linen bleachers and other industrialists.

It should be remembered that at all times the body controlling the waterway remained solely a canal-owning corporation as distinct from a carrying company. In addition to the toll of about 9½d per ton, the freight rate from Belfast to Lough Neagh was

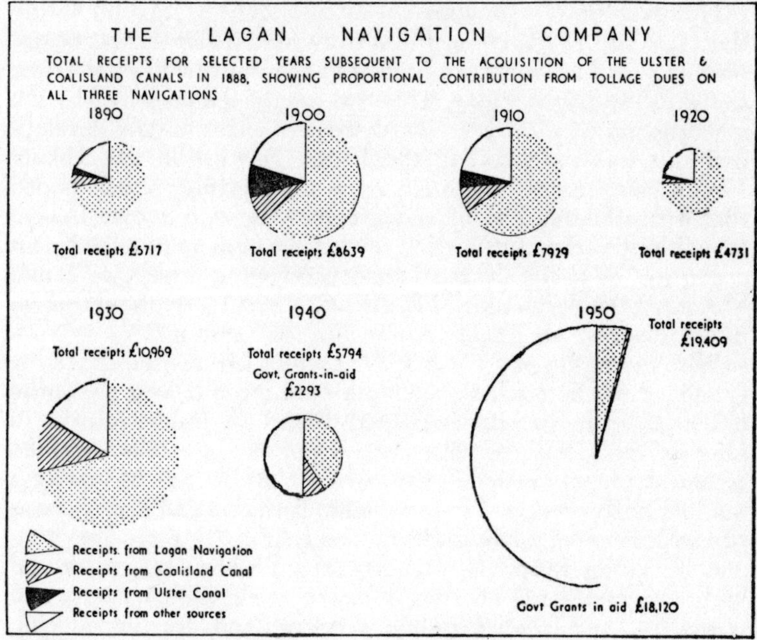

7. Pie chart: Receipts of Lagan Navigation Company

between 5s and 6s. Freight rates to various other landing places were as follows:

	per ton
From Belfast to Lisburn	3s
From Belfast to Boyle's Basin	3s 6d
From Belfast to Hillsborough	3s
From Belfast to Ellis' Gut	4s to 4s 6d
From Belfast to Portadown	6s 6d to 7s
From Belfast to Coalisland	6s 6d to 7s
From Belfast to Newry	6s 6d to 7s

Many of the larger firms provided their own barges and horses; small concerns relied on the services of one of several of the freight-carrying companies operating from Belfast.[33] There were six lighters owned and operated by Lisburn firms while James McCleery of Hanover Quay, Belfast, owned seven, maintaining a regular goods service between Belfast and Moy, Portadown, Blackwatertown and Coalisland.

However, despite the apparent success of the navigation during the 1820s and 1830s, competition from road transport was beginning to constitute a very real threat to the short-haul water-borne traffic in the Lagan valley. This was one of the reasons why the construction of the Ulster Canal seemed such a timely development to those controlling the Lagan Navigation. By linking Lough Neagh with Lough Erne and the Shannon, it was hoped that a continuous line of water communication across Ireland would lead to the introduction of cheaper rates and a consequent expansion of traffic on the Lagan Navigation where, as James McCleery, the Secretary and Registrar, observed 'by the shortness of our canal we are greatly opposed by land carriage'.[34]

With the expiry of the Act of 1814 which had re-established the Company of Undertakers, attempts were made to vest the entire undertaking in the newly-created Board of Public Works in Dublin. The Board, which consisted initially of carefully chosen technical and professional men, was set up in 1831 to replace a number of separate bodies in the administration and management of a wide range of public duties. It was responsible for maintaining and operating hospitals, prisons, inland waterways, piers and harbours, mines and quarries, schools, roads, light railways and tramways, and similar public services, and for introducing measures to relieve famine or social distress. Private interests applied to the Board for financial assistance in carrying out schemes of extension or improvement, and the Government of Westminster usually accepted its recommendations. In the case of inland waterways, the Board replaced the Directors General of Inland Navigation; in the north it took control of the Tyrone Navigation in 1831 and of the Ulster Canal in 1851, and was also responsible for the works of drainage and navigation improvement carried out in the Lough Neagh basin between 1847 and 1859.

A Select Parliamentary Committee met to consider the position,[35] but after protracted discussion agreed that the waterway should remain in private hands, provided it was maintained in a satisfactory condition. So keen had local interests been to keep the waterway free of Government control that they had agreed to spend some £12,000 on immediate improvements over a five-year period. The prospect of controlling a vital link in the inland water communication across Ireland and the benefits to be reaped from contacts with an extensive area lying to the south-west, which the Ulster Canal, then well advanced, would permit, probably contri-

buted to this enthusiasm. The old company was dissolved and replaced by the Lagan Navigation Company, incorporated in 1843[36]
with capital reserves of over £80,000, the price of independence
being a much stricter degree of supervision and control by the
Board of Public Works, at whose insistence a number of restric-
tive clauses were inserted in the Act.

Under the new company the canal competed successfully with
the Ulster Railway, extended as far as Portadown by September
1842, and the new roads of the Lagan valley, particularly those
between Belfast and Lisburn and Lisburn and Hillsborough.[37]
Despite a certain amount of competition in lighter merchandise
from both road and rail, the canal continued to carry a considerable volume of heavy goods, sand, native timber, fireclay goods,
bricks, turf, and a little coal on the outward run, grain (bran and
Indian corn), flour, chemicals, rock-salt, building materials, and
bulky foodstuffs on the inward. Between 1843 and 1859 no dividend was paid to shareholders, as the entire income during these
years was spent on the repair, maintenance, and improvement of
the canal, but neither was any money borrowed from the Government during this period.

By the end of the 1850s the Lagan Navigation Company had
successfully weathered the difficult years following directly on the
transfer of 1843, had fulfilled its obligations to the Board of Public
Works and was in a position to declare a small but steadily increasing dividend, first recommended at 1 per cent in 1859. The
canal was in a stable and reasonably prosperous condition, tolls
and traffic were maintained at a satisfactory level and while there
had been no great increase since the establishing of the Navigation
Company, the period of direct surveillance by the Board of Works
was at an end and the future could be looked to with fair confidence. Traffic was hindered neither by deficiency of water supply
in the summer months nor by ice in winter and as the canal continued to enjoy the traffic created by the construction of lines of
rail communication in and around the Lough Neagh basin without, as yet, the drawback of competition from that quarter, de-
clared dividends rose to 2 per cent. The rapid expansion of railways was soon to tighten the vice of competition but as yet the
navigation could maintain its position as one of the only two commercially successful inland waterways in the north of Ireland.
Downstream traffic in coal had by 1860 almost entirely disappeared and there was already a sizeable import from Belfast and
Newry to the turfless countryside of the Lagan valley.

Reaffirmed in their control of the waterway in 1873,[38] the Navigation Company continued to expand their services and to maintain the channel and works in a most satisfactory condition. Traffic and receipts on the canal during the years 1875–80 were as follows:

Year	Total traffic (tons)	Total Receipts[39]		
		£	s	d
1875–6	90,435	3,917	6	9
1876–7	92,396	3,928	8	9
1877–8	100,203	4,588	4	2
1878–9	105,641	4,242	15	9
1879–80	98,629	4,943	12	6

Total traffic on the waterway had therefore more than doubled since the navigation had passed into the hands of a private company, for in 1836 the estimated figure had been 44,700 tons yielding tolls of approximately £2,000.[40] However, whereas this 44,700 tons had represented about one-eighth of that entering and leaving the port of Belfast at that time, the canal's relative importance had since declined in face of increasing road and rail competition.

A comparison of road and railway rates in operation in 1882 with those current on the parallel and competing navigation is both interesting and significant.[41] The Lagan Navigation Company charged 7½d toll per ton of coal from Belfast to Lisburn while lighter owners charged 2s 2d per ton carriage, including the toll. The railway rate for carriage from Belfast to Lisburn was also 2s 2d but the canal enjoyed an important advantage in this trade because the coal could be delivered by lighter direct to the mills and bleaching works of Lisburn and district, most of which had been built either before the advent of the railway or at least before it had gained sufficient momentum to exert a strong pull on the siting of industrial concerns. On the other hand, break of bulk and costly road haulage were required if coal was dispatched by rail, which mill owners were anxious to avoid if at all possible because of damage to the coal.

The lighter rate to Kinnego Gut or Port Lurgan, as it was sometimes called, was 3s 3d, the railway rate 3s 9d, while that to Portadown was 3s 9d by water for both coal and grain. Out of the 2s 2d per ton charged for lighter carriage between Belfast and Lisburn, the Lagan Navigation Company received 7½d and the carriers the remainder; out of the 3s 3d to Kinnego they received 1s 0d, out of the 3s 9d to Portadown they received 1s 1½d, the same to Coal-island. There was an extra charge for the use of a tug, the rate to Portadown being 14s 0d for a train of up to four lighters, each carrying from 60 to 90 tons. When one compares these figures with those in operation half a century earlier, in the 1830s, the liberal nature of the controlling company, combined with the effect of increased competition from other forms of transport, particularly the railways, can be seen.

Traffic on the Lagan Navigation in 1836[42]

Journey	No. of Boats	Total Tonnage	Rates of Tolls			Amount of Tolls		
			£	s	d	£	s	d
Belfast to Lough Neagh	370	18,500	3	0	0	1,110	0	0
Belfast to the Summit Level	204	10,200	2	0	0	408	0	0
Belfast to Lisburn	197	9,850	1	10	0	295	10	0
Belfast to Drumbridge	—	—	1	0	0	—		
Lough Neagh to Belfast	77	3,850	2	10	0	192	10	0
Summit Level to Belfast	32	1,600	1	14	1	54	10	8
Intermediate Places	14	700	—			—		
	894	44,700				£2,060	10	8

Broadly speaking, the canal rates and tariffs current in the 1870s and 1880s compared quite favourably with those in operation on the competing railways. The main disadvantage of the waterways was slowness of traffic but they still continued to handle a considerable proportion of the commerce in heavy, imperishable goods, grain and fuel, timber and machinery, in which speed of delivery was of less importance than an even, uninterrupted journey and whose bulk made them better suited to take advantage of the lower tariffs in existence on the various inland navigations.

At this time the state of the Ulster Canal was of serious concern to the Lagan Navigation Company. Canal freight rates to places like Benburb, Monaghan, Clones and Belturbet were slightly cheaper than by rail. Moreover, the canal was often more convenient, because some railway stations were over a mile from the town centre, making road carriage necessary. Although 30,000 tons a year were passing up the Lagan to Lough Neagh, almost no traffic was being exchanged between the Lagan and the Ulster line to Upper Lough Erne and Enniskillen, because of the need for transhipment to and from craft that could work over the shallower Ulster Canal with its narrower locks. As a result the trade of much of Counties Armagh, Monaghan, Cavan and Fermanagh went by road or rail. Since 1865 the Ulster Canal had been controlled by the Board of Public Works, who had spent £20,000 upon it without increasing traffic. Representatives of the Board, however, stated that the Government would make further improvements to the Ulster line to permit substantial and regular use if special lighters were put on it which could work through to Belfast. The Lagan company therefore built five such craft between 1878 and 1880, hoping not only to encourage the through trade, but to increase traffic on the Lagan itself.

The flourishing Lagan Navigation Company was eventually induced to take over the Ulster Canal, and also the Tyrone Navigation, by a private Act of 1888.[43] This bound the company to spend over £10,000 on the Ulster Canal during the next few years, but even after further major repairs, costing £12,700, the canal remained unfit for traffic for most of the year, especially the western section beyond Clones. Not only had the Lagan Navigation Company been saddled with a waterway which no amount of money could have rendered properly navigable, but it had also to spend a great deal more on useless repair work, besides paying for the Act of 1888 by which the millstone was adjusted around its neck.

The Lagan Navigation itself remained in very good order, and traffic continued to expand as the nineteenth century drew to its close. In the year ending 31 March 1893, for example, traffic totalled some 153,000 tons, of which four-fifths was upstream, and net tolls during the same period amounted to £4,800, an average toll per ton of 7¾d;[44] by 1898 traffic had risen to over 170,000 tons and net tollage receipt to £5,600. Traffic consisted mainly of coal, grain, and general merchandise upstream from Belfast and sand, native timber, fireclay goods, and bricks downstream towards the city. The average haul per ton was nineteen miles for upstream cargoes and twenty-five miles for downstream cargoes, with coal representing about two-thirds of the total traffic, all of it moving upstream.

These figures must be set against the substantial obligations with which the company was now confronted. Not only was it still committed to an annual rent to the Government of £300*, a

---

* Under the terms of the Act of 1843 this annual rent was not levied until the financial year ending 31 March 1848, presumably to give the new company time to become more firmly established in control of the canal. In any case the Act of 1843 bound the company to spend over £12,000 on immediate improvements over a five-year term, the annual levy becoming operative at the end of this period.

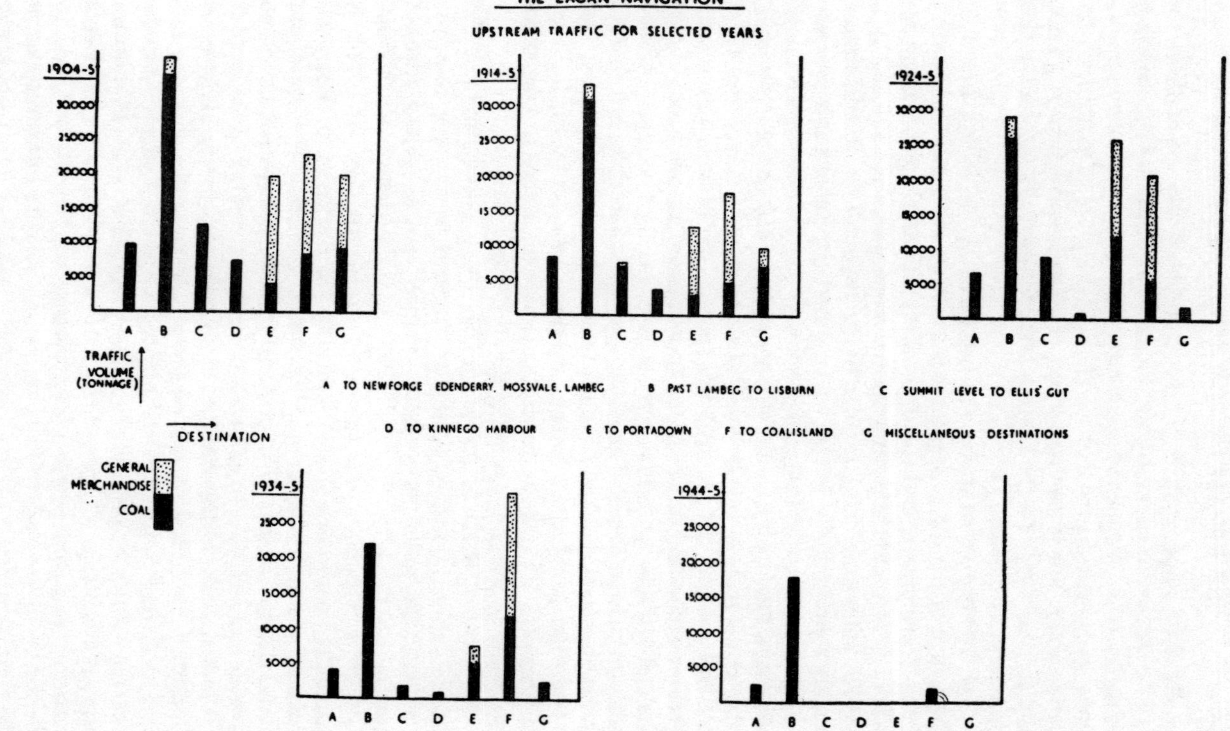

8. Bar chart: Lagan Navigation upstream traffic

yearly levy apparently in consideration of the expenditure of public money, as raised by local taxation, during the construction of the canal in the previous century, and one of the many stipulations laid down in the Act of 1843 as the price of continued independence, but it was now compelled to spend between £1,000 and £1,500 a year on the Ulster Canal, whose tolls and rent at this time never exceeded £800!

The upstream coal traffic was halved by World War I, but with the return of the canal to private ownership after wartime control from Westminster, and the expenditure of nearly £20,000 on repairs and renewals, the navigation was once more in a position to handle substantial and regular barge traffic. The decade immediately following the setting up of Northern Ireland as a separate political unit was its last really successful period. For seven out of ten years, total traffic on the waterway exceeded 100,000 tons and, with the increases in tolls which had occurred in 1916 and 1920, net toll receipts often exceeded £8,000 per annum. After coal, grain remained by far the most important commodity moving up the canal, maize or Indian corn, which was then brought in huge grain vessels to the Belfast docks and loaded directly into barges moored alongside. Each lighter, containing 80 to 85 tons, was horse-drawn to Lough Neagh, although by now motor lighters, independent of tractive power on lough or river, had been introduced by many of the more enterprising firms.

From an analysis of traffic on the Lagan Navigation for the 1920s (Fig. 8) it can be seen that upwards to Lisburn coal was predominant, totalling 307,033 tons as against 19,126 tons of general merchandise. Beyond Lisburn to the summit level, Ellis' Gut and Kinnego, the relatively small tonnage involved is similarly composed. To Portadown the tonnages of coal and general merchandise arriving by water from Belfast were about equal but to Coalisland imports of merchandise totalled 127,876 tons compared with 60,186 tons of coal, due mainly to the import of Indian corn to the mills at the Coalisland basin. Here canal transport continued to compete quite successfully with the circuitous Great Northern Railway branch from Dungannon, the two main mills being nearer the canal basin but at some distance from the railway. There was also still a small movement of coal inwards from the Lagan Navigation along the Ulster Canal to Benburg, till its closure in 1931. From this date onwards, except for grain imports to Portadown and Coalisland, upstream traffic on the canal consisted almost entirely of coal.

Over the same period the main downstream traffic was washed sand and gravel for building purposes in Belfast, exported in large quantities from the Lough Neagh shores of east Tyrone. The rest was native timber, peat moss, fireclay goods, bricks and miscellaneous agricultural produce and manufactured goods from the towns served by the Ulster Canal, Coalisland, the Lough Neagh ports and Portadown. Traffic entering the Lagan Navigation from the Ulster Canal in the 1920s was very fluctuating, depending largely on the water supply of a particular season and the general current of public opinion upon the relative merits of canal, road and railway.

The years immediately before the outbreak of World War II saw a sudden decline in the traffic, receipts and comparative importance of the last inland canal in Northern Ireland to maintain a position of independent commercial significance. The slowness of canal transport was thought to be inherently incompatible with modern commerce and its emphasis on speed of delivery. Though heavy goods continued to move up and down the Lagan corridor stretching south-westwards from Belfast, between producer and consumer, between point of import and point of requirement, they were carried by the railways and to an increasing extent by the roads. Once begun, the decline in the use made of the Lagan Navigation, and to a greater extent of the other inland waterways in the north of Ireland still open for traffic, continued throughout the remainder of their existence. Total downstream traffic on the Lagan declined from over 29,000 tons for the twelve months ending 31 March 1931 to a mere 600 tons in the corresponding period ten years later and upstream traffic from 87,000 tons to 28,000 tons over the same period.

With the drastic reduction in toll receipts that accompanied this decline in traffic, the company's funds were completely exhausted by the late 1930s and, failing substantial financial help from the Government, the closure of both the Lagan and Tyrone Navigations appeared imminent. The Government, however, were particularly anxious that the two waterways should remain open, for should they fall into a state of disrepair which precluded the passage of commercial traffic, by S. 222 of the Act of 1843 the Lagan Navigation Company would automatically cease to exist and, as successors in Northern Ireland to the Board of Public Works, control of the canals would pass to the Ministry of Commerce. Rather than assume responsibility for a commercial undertaking already moribund, the Government of Northern Ireland agreed to subsidize the Lagan Navigation Company and so enable it to remain in nominal control of the Lagan and Tyrone Navigations. Beginning in July 1938, abstracts of monthly receipts and payments taken from the company's cash book were sent to the Ministry of Commerce who then sent a cheque to make good the deficiency. These grants-in-aid were continued during the remainder of the waterway's existence, not only to make good the normal deficiency due to maintenance but also to pay for various major works of repair and replacement authorized from time to time. Between 1938 and 1952 over £130,000 was paid to the Lagan Navigation Company from public funds.

Despite these subsidies the end of the navigation as a commercial waterway was not far away. The canal had so declined in importance that it was little affected by World War II, and in the post-war years there remained only a negligible traffic in military stores to and from the Thiepval Barracks at Lisburn and a slight downstream trickle from Newforge between 1948 and 1950. After 1947 there was no traffic in either direction above Lisburn. Eventually, in December 1953, the Minister of Commerce, Lord Glentoran, announced his Ministry's intention to abandon the Lagan, Upper Bann, and Coalisland Navigations in the following spring, stating that they were no longer an economic means of transport and that they had outlived their usefulness, even if restored and fully developed. He proposed abandonment for all three, except for the section of the Lagan Canal between Stranmillis and Lisburn, drawing attention to the fact that some £140,000 had been paid to the Lagan Navigation Company since 1937 out of public funds in face of declining traffic and rising costs of maintenance and repair, and that such liberal subsidies could not continue indefinitely.

By the Inland Navigation Act (N.I.) of April 1954, the Lagan Navigation Company was dissolved and the canal between the Union Locks and Lough Neagh was officially abandoned for all navigational purposes. The Act provided that the Sprucefield-Stranmillis section should, for the time being, be kept open for traffic with a rider permitting subsequent abandonment, if necessary. The upper reaches of the Lagan Canal, the Tyrone Navigation and that portion of the Upper Bann Navigation comprising the lower courses of the Upper Bann and Blackwater rivers, all of which were now officially abandoned, were to be made safe and maintained so that they might continue to discharge surface drainage water, while in the case of subsequent closure of the lower Lagan Navigation, the necessary water levels for industry were to be maintained. By the same Act control of the various waterways finally passed to the Ministry of Commerce in the Government of Northern Ireland.

After April 1954, therefore, the Ministry had complete control of the Lagan Navigation[45] but only the reach between the Union Locks and Belfast still carried any traffic. The final move in the long and complex history of the navigation occurred in the spring of 1958 when the Ministry announced its intention of making an order for the abandonment of the lower reaches of the canal, between the first locks at Stranmillis and the Union Locks at Sprucefield, near Lisburn, from 1 July of that year. The canal had not been used by boats for some time and the final closure took place smoothly and quietly, unnoticed by the general public for whom the waterway then had but little significance.

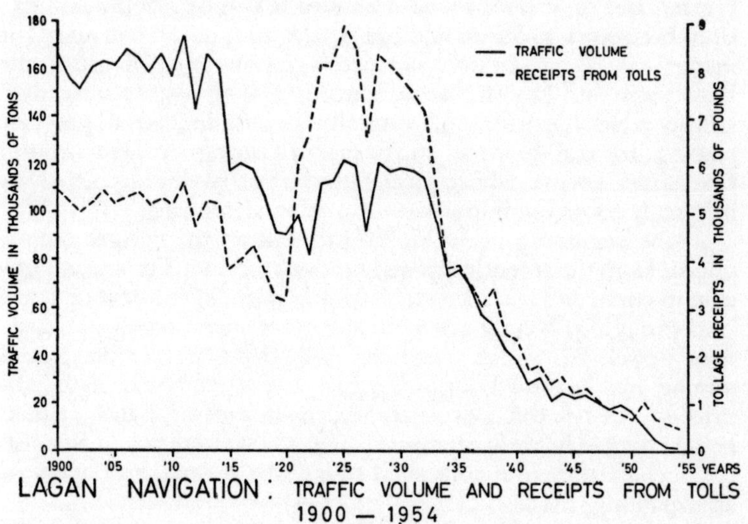

10. Lagan Navigation graph

CHAPTER IV

The Tyrone Navigation
(Coalisland and Ducart's Canals)

ASSOCIATED closely with the history of coal working in east
Tyrone was the development of a canal system to transport it
cheaply and efficiently to the nearest port. This system, begun in
the early eighteenth century, developed as coal production increased and only since the end of World War II has the last section, between Coalisland and the Blackwater, become disused for
navigation, to remain only as a drainage channel feeding into the
Blackwater and Lough Neagh. During its long history it handled
considerable quantities of agricultural and industrial produce
passing to and from a small manufacturing enclave around
Coalisland, an area whose precocious industrial development was
originally rooted in important local mineral deposits.

At the beginning of the eighteenth century the meagre output
of coal from the recently-opened workings in east Tyrone was unable to command a market in Dublin because of the cost of over-
land cartage to Newry and subsequent shipment south by coastwise vessel. To cover transport costs from the remote inland
mining area beyond Lough Neagh to the consumer in Dublin, a
price of 18s per ton was necessary, much greater than that then
being charged in the Irish capital for the best English and Scottish
coal, which therefore continued to enjoy a relative monopoly of
an expanding market.

Though not reaching the height of its powers until later in the
century, already by the 1730s Dublin was a cosmopolitan centre
renowned for the brilliance of its intellects and the wealth of its
gentry, a city in which social and economic conditions bore little
relation to those permeating society throughout a greater part of
the country. To the capital for at least the winter months came the
great majority of those with social aspirations of one kind or

another and it was amongst these gentry and aristocracy that the new fuel aroused a curiosity which in turn created a slight but increasing demand.

As most of the early coal workings in east Tyrone were in the hands of Dublin interests, it was natural that the ease, speed and cheapness with which coal could be brought to Newry, the sea-port nearest to the capital, should be of paramount importance; to them the desirability of a linking waterway was undeniable. In 1709, some fifteen years after the recorded discovery of coal seams at Drumglass, south-west of Coalisland, a petition was presented to the Irish Parliament[1] by Thomas Knox, a colliery proprietor, outlining the advantages of cutting a canal—a precursor of the Newry Navigation—from Knock Bridge, near Gilford, on the Upper Bann, to Fathom Point, below Newry, by which coal from Tyrone might be conveyed from Lough Neagh to Newry *en route* for Dublin, but although a Committee of the Irish House of Commons reported favourably on the proposal, nothing was achieved and support for the idea evaporated.

Later, as the advantages of a productive Irish coalfield and the need for adequate coal stocks to be held in Dublin were realized, influential voices were raised to advocate a canal to link Dungannon, near Coalisland, with Dublin. Thus, in 1727 Thomas Prior wrote to urge canal construction south-east and south-west of Lough Neagh, by which 'we may be furnished with coals of our own country for our own consumption',[2] to replace an estimated annual import of between 60,000 and 70,000 tons, while in 1729 Arthur Dobbs, the Surveyor-General, regarded the building of a canal from the pits to Lough Neagh as relatively easy and emphasized the advantages when 'the coals of Tyrone are being daily found better and in increasing quantities'.[3] In this same year an interesting pamphlet, *Remarks on a Scheme for supplying Dublin with Coals*, was published in Belfast.[4] The author, Francis Seymour, owned a pit at Brackaville near Coalisland. He strongly supported Knox's proposals of 1709, and also remarks that were a cut to be made through a bog which then extended from 'the foot of Drumglass coal hill to a river [the Torrent] which enters itself into the Blackwater', the fuel could be conveniently carried in small lighters to the 'gablards' or other craft on the lough, and thence to Newry.

In 1729, also, the Commissioners of Inland Navigation for Ireland were established. They studied the proposals that had been made, and in 1732 authorized the construction of a canal, one year after they had agreed to the building of the complementary Newry Navigation. The plan was to make a cut from Coalisland to the Blackwater, some 4½ miles to the south-east, down which might pass lighters linking up with craft on Lough Neagh, which could then ply to the port of Newry.

Work was set in hand in the summer of 1733 under the general supervision of Acheson Johnston, but progress was extremely slow. The fact that the head of the canal at Coalisland was some way from most of the early coal-workings at Drumglass in the south-western extremity of the area of coal measure outcrop proved a serious disadvantage. As the canal lengthened eastwards, land carriage from the pits still kept transport costs well above what they would otherwise have been and in 1753[5] a petition was presented in the Irish Parliament by a company established in 1749 with an initial capital of £10,000 and headed by the Archbishop of Armagh, the Archbishop of Tuam, the Rev. Hon. Arthur Hill, Charles Caulfeild and Thomas Staples, which sought governmental help in building a suitable roadway, three miles in length from Drumglass to Coalisland. By such a connection, they said, repeating earlier arguments, the export of Tyrone coal would be made easier and the undesirable dependence of the Irish market on imports from Irvine, Saltcoats, Whitehaven, Workington, Flint, and Swansea, in all about 70,000 tons per annum, would be greatly reduced. South-west gales and keen frosts often prevented ships from leaving British ports, supplies of coal were not always reliable and it seemed desirable to create a stock-pile of fuel in Dublin from local deposits. This would be a simple matter with the construction of a navigable waterway from Coalisland to Lough Neagh, and with a ship canal providing improved access to the port of Newry,[6] and £4,000 was granted shortly afterwards for the construction of the proposed roadway from Coalisland to Drumglass.

Meanwhile the canal below Coalisland was being built very slowly. Criticisms were made[7] that it was useless to sink new pits and develop underground coal workings while the new cut remained incomplete, but the general fervour attending the expansion of mining and an increasing coal output were more than sufficient to outweigh disappointment at the tardy progress on the canal. By the 1750s, indeed, coal was being won in quantities quite sufficient to supply all Dublin's requirements, and with the anticipated new roadway and the completion of the waterway to the Blackwater, it was estimated that coal would soon be shipped

VII. Lagan Navigation: (*above*) the Union Locks, Sprucefield, with early nineteenth-century lock-house on the left; (*below*) the aqueduct over the river Lagan near Moira, built by Richard Owen between 1782 and 1785

VIII. Lagan Navigation: (*above*) road bridge at Ballyskeagh near Lambeg; (*below*) the former canal wharf and stores at Lisburn

IX. Tyrone Navigation: (*above*) the Moor or McAliskey's lock (No. 5) on the Coalisland Canal in 1945. It bears the date 1749 and Acheson Johnston's name; (*below*) the canal basin and mill (*left*) at Coalisland in 1945

X. Tyrone Navigation: (*above*) Mack's bridge over the staircase (double) lock on the Coalisland Canal; (*below*) Newmills aqueduct on Ducart's Canal

from Coalisland at 6s per ton, a price which would enable it to break the British monopoly on the Dublin market.

Work continued slowly on the short four-mile cut eastwards from Coalisland during the 1750s and 1760s. At the head of the navigation, at 'Coal-Island' itself, an extensive basin was constructed, into which a feeder from the Torrent, the main left bank tributary of the Blackwater, poured the bulk of the canal's water supply. Unfortunately the Torrent, an ungraded, powerful mountain stream, brought down considerable quantities of silt, clay and stones and during the latter half of the eighteenth century these deposits went a long way to rendering useless not only the greater part of the basin but also several of the locks and levels farther downstream. The first and second locks had to be constructed on running sand, and lock construction should therefore have been preceded by the sinking of large numbers of piles, while the fifth, sixth, and seventh locks, towards the Blackwater, were sunk in equally unsuitable peat bog, again without adequate piling and paving of the lock chamber floors. Both these failings were soon to have disastrous results.

Furthermore, along several of the levels springs fed into the canal through the sandy banks and caused frequent flooding, while between the sixth and seventh lock the canal bed was cut through 'a very low swampy turf mould' with the Torrent close to its south bank and another stream on the north, separated 'only by a very narrow spongy trackway or rampart'. In this section of its course, as might be expected, the parallel streams flooded the canal when water was least needed and drained it in dry weather. The unsuitable nature of the surface deposits for large-scale excavation and lock construction allied to inexperience in the choice of feeders, precise siting of locks and difficulties experienced in joining the canal to the Blackwater, presented problems in the building of the Tyrone Navigation that were much greater than those encountered on the much more extensive Newry Canal, which took ten years to finish, and engineering works dragged on for the greater part of the century. Between 1746 and 1782 the Coalisland Canal received some £25,000 from public funds, while between 1753 and 1767 the canal works and various ambitious mining schemes in the area received over £112,000.[8]

To appreciate fully the reasons behind this unbridled expenditure of public money it is necessary to examine the general state of public finances in Ireland at the time. In the eighteenth century the basis of the Irish revenue law was certain legislation of Charles

E

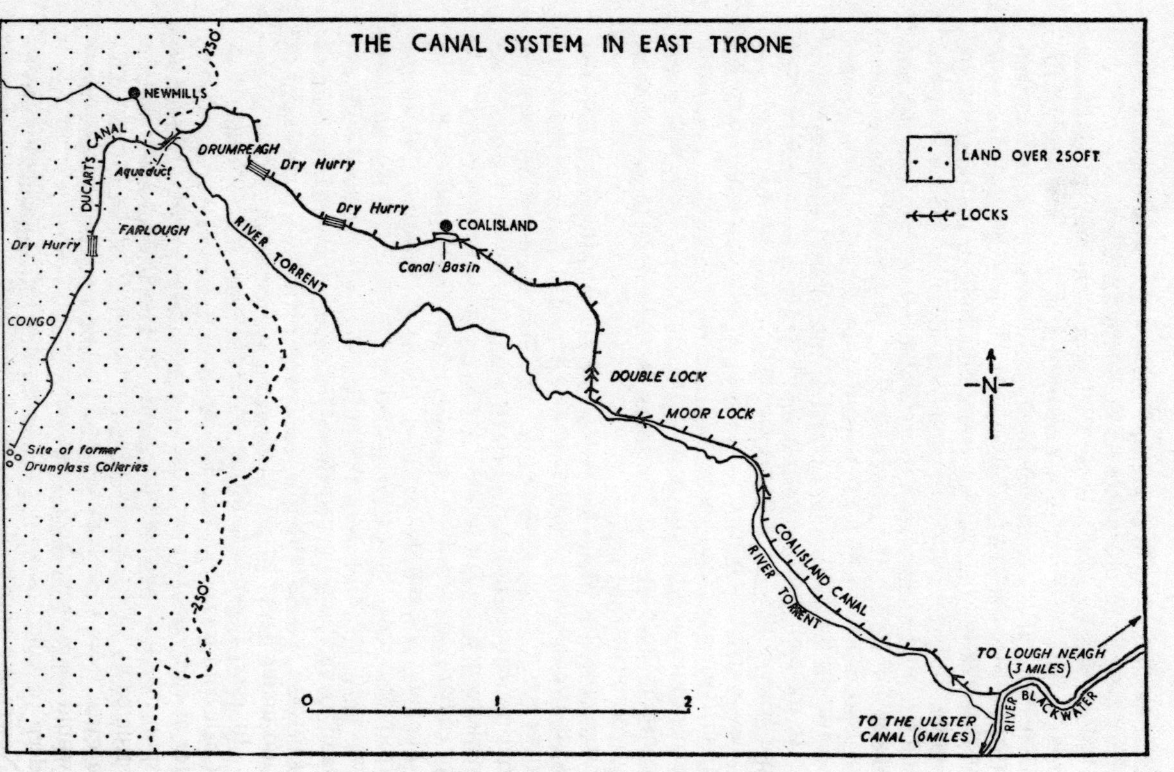

9. Map of canals in East Tyrone

II, which followed on the confiscations after the rebellion of 1641 and which provided that in return for the King's surrender of his right to the full benefit of these forfeitures, he should be granted a legal hereditary revenue. The older forms of Crown property were abolished and the new revenue then granted was vested for-ever in the King and his successors. The most important sources of this revenue were, firstly, the Crown rents arising from the land confiscations of Henry VIII and from the counties forfeited after the Earl of Tyrone's rebellion in the reign of Elizabeth I; secondly, the quit rents arising out of property forfeited after the rebellion of 1641; thirdly, the 'hearth money'; and fourthly, the excise and customs duties and licenses for selling ale, wine, and spirits.

In the eighteenth century a number of attempts were made by the Irish Parliament to gain control of this hereditary revenue. These were all unsuccessful but in order to redress the balance the Irish Parliament had recourse to a plan of keeping local public expenditure as high as possible, thus compelling the King to make constant application to Parliament to maintain some semblance of governmental control over Ireland. From 1753 onwards, after an acute constitutional struggle, it became the custom to grant large sums every session, nominally for public works such as canals and river navigations, bridges, and the encouragement of industry, but in many cases for the private aggrandisement of members of parliament and their friends. There was an additional inducement to embark on canal building schemes, profitable and unprofitable alike, and the full development of the Irish canal system stems from this time. Between 1730 and 1787, for instance, £351,946 was used out of the tillage dues for the purpose of furthering in-land navigation in various parts of Ireland and in addition £505,436 was granted from parliamentary sources, local duties, and under Royal letters.

In east Tyrone the proposed roadway from Drumglass to Coalisland was either never completed or did not prove a success, for as early as 1 January 1760 Thomas Omer, who had been re-sponsible for much of the work on the lower Lagan, was in-structed by a Parliamentary Committee to survey and report upon a canal from the Newry Navigation to the Tyrone collieries. It would seem that Omer was instructed to consider a canal 9½ ft deep to take 100 ton craft, presumably with the idea that the Newry ship canal could later be connected with it by an enlargement of the Newry Navigation, so that coastal vessels could pass directly to the collieries. Omer later recorded[9] 'that he was of Opinion it would not answer, but made an Estimate'. He himself seems to have proposed as an alternative a canal from the Coalisland basin by way of Creenagh, 3 miles 159 yd long with 16 locks, at a cost of £15,667 10s. In 1761 the Committee reported that 'it would be of great advantage to carry the canal higher up into County Tyrone from the [Coalisland] Basin to Fairlough and, this being done, a sufficiency of coals will be supplied to Dublin without dependence on a foreign market'.[10] Coal imports having now increased from some 70,000 tons a year to about 180,000,[11] Parliament authorized the extension, but on the full scale for 100-ton craft. Thomas Omer began to carry it out at the same time as he was at work on the ship canal at Newry, but handed over in June 1762 to Christopher Myers, a British architectural engineer employed by the Office of Public Works who worked with Omer on the ship canal after John Golborne's dismissal.

The plan was to make navigable the Torrent river as far as possible, and then to build a canal to near Congo. Myers built half a mile of navigable waterway upstream from Coalisland, 45 ft broad at top, 22 ft at bottom, and 10 ft deep, and also a partially-completed lock, 125 ft long and 22 ft wide. Then, in the light of his experience, he told Parliament how much the whole canal would cost, and added that in his view it was 'impracticable to make a Navigation up the Bed of the River, as it was first intended, or if it was practicable, the Time and Expense necessary for such an extensive Undertaking would be more than equal to the Utility of it when finished'.[12] Instead, he recommended a canal of the same size as the Newry Navigation, to take craft 60 ft long, 12 ft broad and drawing 3 ft, to carry between 50 and 60 tons, and thought the whole line from the collieries to Newry ought to be built 6 ft deep to provide an ample margin for craft which, as well as using the canals, could navigate Lough Neagh, and make coastwise voyages, at least in the summer. For such a canal, with fourteen locks, from the Coalisland basin through the townlands of Derry and Drumreagh, over the river Torrent by an aqueduct, and so to Farlough, his estimate was £18,141.

In 1767 Parliament agreed to adopt Myers's proposal, and voted £5,000 towards it. However, a second opinion was sought on the engineering involved and Myers's plan was superseded by that of Daviso de Arcort[13] (Ducart), an architect and engineer of mixed Franco-Italian descent who had made his name 'in the Sardinian service', and who later designed the Custom House at Limerick, Castletown House, Co. Kilkenny, and Kilshanwick House, Co.

Cork. Ducart favoured an entirely new scheme which in two respects owed much to the Duke of Bridgewater. The Bridgewater Canal had reached the Manchester terminus at Castlefield in 1765, and there, to get cargoes up to the Deansgate level, Brindley built a shaft beside the end of the canal in a sandstone tunnel. A water-powered crane over the shaft hauled up boxes of coal from the boats.[14] Construction was actually in progress in 1765, and the plan was publicly known. At the other end of the canal, at Worsley, the surface canal connected on the level with small waterways carrying 12-ton boats that ran into the mine, and which served the double purpose of enabling coal to be boated out, and of drainage. Later, the same system of hauling boxes of coal from one level to another of these underground canals was used, but probably not till after that at Castlefield had been built.[15]

It is more than likely that Ducart knew of these developments when he proposed that a small or tub-boat canal just over 4 miles long, 24 ft wide and 4 ft deep, should be built to the collieries, the boats on which, like the Duke's, would carry 12 tons. This he planned to run from the 'old Navigation' near Coalisland partly in the open and partly in a tunnel that would be both a navigation and a drainage sough or level, to Derry colliery. There a shaft 30 ft in diameter and 148 ft high, with walls 3 ft 6 in. thick, was to be built. Coal was to be let down this in 10 cwt boxes from the colliery workings, and from the upper canal level leading on to Drumglass, by means of a capstan, the shaft also being provided with two spiral staircases 10 ft wide. From the top of the shaft a further length of canal, partly in the open, and partly in a second tunnel was to lead on the level to Drumglass, crossing the river Torrent over a three-arched aqueduct at Newmills. The tunnel sections were to be provided with passing places and ventilating shafts.

It would appear that Ducart's proposals were accepted by the Irish Parliament for by November 1767 he had spent £3,839, and had built about a quarter of a mile of finished canal and half a mile of partly completed work between Farlough and Derry on the upper pound, together with part of the aqueduct. He considered £14,457 more would be needed to complete the works. By February 1768, however, after much discussion before the Parliamentary Committee by those expert and not so expert he had revised his estimate to £26,802, exclusive of the salaries of an engineer and two assistants for two years.[16]

Once again, however, the plans were changed. The section of canal already started was retained, and the fine Newmills aqueduct of beautifully shaped freestone was completed. Instead of the shaft at Derry colliery, however, the line was altered to provide an inclined plane[17] (called a 'hurry' or 'dry wherry' at the time) at Drumreagh House with a fall of 65 ft, and another, with a fall of 55 ft, 844 yd to the south-east close to what was marked on contemporary maps as the 'Great Water Wheel'. Another 942 yd of canal led to the Coalisland basin. The pound between Drumreagh and Farlough was 3,034 yd long. Upwards from Farlough the idea of tunnelling was dropped, and a third 'hurry' was built just west of Farlough lake to raise the line 70 ft in a horizontal distance of 119 yd to its summit level, whence the line of canal ran south-westwards for 3,080 yd to the Drumglass collieries.[18] Before this final line was settled, there had been a scheme to build the canal only from Drumglass to the aqueduct, where boats would be unloaded into small lighters moored directly beneath on the Torrent, but this was not considered practicable.

By the summer of 1773, it seems that the canal was completed, but that the inclined planes had been only partially built, and that work on them had been stopped pending a visit from either John Smeaton or his assistant William Jessop, both of whom were in Ireland inspecting the proposed route of the Grand Canal from Dublin to the Shannon. Smeaton had worked for many years with William's father, Josias, on the Eddystone lighthouse, and after his death in 1761, became guardian to the young William, then 16, who worked as his pupil and later as his assistant. Smeaton did not come north to Tyrone in 1773 but sent young Jessop, for whom he had a high regard, to inspect the canal works there and to report back to him on his return to England. Unfortunately, a copy of Jessop's report cannot be traced, but we have Smeaton's views upon it, written from Yorkshire in April 1774, without himself visiting the canal.

It is probable that Ducart had begun to construct timber ramps laid on the causeways of the inclined planes, the ramps being fitted with rollers. This device may well have been an adaptation of the *ponts aux rouleaux* described in a work published in Paris in 1693, at which time examples powered by waterwheels are said to have been in use successfully for some time in Holland. However, Ducart experienced considerable difficulty in providing the power necessary to pull his one-ton boats up the ramps and engineering work came to a halt.

Smeaton suggested that they should be counterbalanced, a loaded boat in descent being made to haul up an empty one by connecting the two with a rope running round a wheel at the top of the plane, that the size of the boats should be increased from one to two tons, and that 'a tackle, assisted by a winch or wind-lass', should be provided to haul down-going boats over the sill of the upper pound before they started down the plane. He put forward these proposals as the best ways he could think of to make the project work, though he made it clear that 'the circumstances attending it are such, that I never could have recommended a canal of any kind, as I am sensible it never could be made in any degree to answer the expense', and preferred a horse railroad.

All three of his suggestions seem to have been adopted. The ramps were built double so that the boats could be counterbalanced, the size of the boats was increased to 10 ft by 4½ ft,[19] working in pairs chained together, and a horse gin was provided at each plane to haul craft out of the upper pound. In practice, however, the planes still did not work very well. 'Finding various inconveniences, from some of the rollers not turning, and from the individual irregularities of the diameters of others throwing his boats to one side, as well as from other causes,'[20] Ducart substituted parallel railways, each of which carried a cradle mounted on four wheels upon which the boat rested. At Coalisland, the canal ended about 15 ft above the main canal basin. Here the boats were again floated on to a cradle and run along a short length of rails 'to the wharf; and there, being supported on geers and frames, the boats were turned over to discharge their cargoes'.[21]

During the work it seems that Smeaton's proposal to substitute railroads along the entire length of Ducart's Canal was considered by a Committee of the Dublin House of Commons, who were given evidence that the work could be done for a reasonable sum, but though the work was put in hand, it was soon abandoned. By 1777 the canal was at last finished,[22] and the first boats moving downstream from the Drumglass collieries successfully negotiated the 'hurry' at Farlough, the Newmills aqueduct, the 'hurries' at Drumreagh and towards Coalisland, and arrived at the canal basin there amid scenes of mild enthusiasm.

These inclined planes were the first canal structures of their kind to be built in the British Isles. They were soon followed by others: one on the Ketley Canal in Shropshire in 1788, which was also counterbalanced, others on the Shropshire and Shrewsbury

Canals, a third inside the Worsley mine near Manchester, in the 1790s, and later, by further examples in the Westcountry, south Wales and elsewhere. They were especially useful in hilly country, where they were a successful substitution for flights of locks which would have been wasteful of both water and time.

In authorizing and financing the construction of Ducart's Canal the Irish Parliament had hoped that traffic on it would not be restricted to coal, and that it would benefit the general trade of a wide area. No through traffic in either coal or merchandise could, however, be handled till the main navigation between Coalisland and the Blackwater had been completed.

Not till 1787 was it at last finished, some fifty-five years after engineering work had been set in hand. A number of factors explain the immense delay: unexpected engineering difficulties, accentuated by the general inexperience of the 1730s and 1740s in dealing with the problems of canal construction; a shift of emphasis at a crucial time, about 1750, from work below Coalisland to consideration of how useful a headward extension above the town would be, and a subsequent diversion of money and manpower in the 1760s; a slowing down of the grants from the Irish Parliament in the 1770s and 1780s; and, perhaps, a temporary loss of faith in east Tyrone's ability to supply Dublin's coal requirements or to replace the substantial coal imports of the time, and so a loss of influential enthusiasm.

At Coalisland 'the bason . . . is a spacious harbour surrounding an Island of an oblong figure. The original intention was that the boats should come up on one side, take in their cargoes at the head of the island, and come down on the other side, and so depart'.[23] Below the town the navigation was larger than Ducart's Canal. It had seven locks with a total fall of 51 ft, a depth of 5 ft, and towing paths on both sides. The water level was maintained by two feeders at the Coalisland basin, one from a small dam, the other from the Torrent.[24]

The Newry Canal having been completed as long ago as 1742, linking that port with the Upper Bann and Lough Neagh, there was now a continuous inland waterway from the area of mining activity to the chief maritime port in north-east Ireland and coal could be extracted in increasing quantities in the knowledge that cheap and efficient water transport between pit head and consumer would keep its selling price at an attractive level.

However, the optimistic anticipations attending the construction of both navigations were never realized. In the case of Ducart's Canal unforeseen difficulties proved disastrous: it was soon found that the gradient of the planes was too steep to allow them to work easily by normal counterbalancing and that a sufficient supply of water could not be retained in the upper reaches of the canal because of limestone seepage. In fact, the laborious construction of the Coalisland-Drumglass extension had been a gross misapplication of effort, the failure of which had probably been realized long before its completion but which had been finished for no other reason than to prove the feasibility of its engineering and to prevent the prosecution of Ducart for blatant misuse of public money. At any rate by 1787, when the Corporation for Inland Navigation sent Richard Owen, engineer of the extension of the Lagan Navigation from Lisburn to Lough Neagh, to inspect the works and report on the condition of the canal above Coalisland, he found the entire section inoperative due both to a serious shortage of water in the westernmost stretches and the mechanical failure of the various 'hurrys'.[25]

He himself proposed that the 'hurrys' should be abolished and 'adequate inclined roads . . . either paved or gravelled' should be substituted. He then proposed to put on each level of Ducart's Canal flat-bottomed craft 40 ft long, 13 ft wide and $2\frac{1}{2}$ ft deep, each of which would carry 14 small four-wheeled wagons to hold one ton of coal each, in two rows, and which would be towed in pairs by one horse. At each incline the wagons would be run off the boats and 'all or part of the Waggons may be taken down together at pleasure by locking some of the Wheels': at Coalisland the wagons would be tipped into the barges.

Owen put forward an alternative solution, a new canal, mostly in tunnel, starting from the Coalisland Canal at the lock below the double lock and running in a curve along the valley of the river Torrent to the line of the Dungannon road, where it would serve the local collieries, heading thence due west to Drumglass. He estimated the total length as about 6,550 yd, of which between 4,400 and 4,960 yd would be in a tunnel 7 ft wide, with a clearance of 5 ft 3 in. to 5 ft 6 in. above water level. Besides acting as a navigable canal this subterranean cut would also serve as a drainage sough or level by which the problem of flooding in the Drumglass pits, already attracting attention (see note 31), might be partly remedied.

His suggestions were not accepted by the Corporation and it would appear that at last Parliament had become impatient with the continued loss of public money, not only in the Coalisland area, that 'happy hunting ground for engineering experiment', but on inland navigation throughout the country as a whole. In 1787 the national body or Corporation was dissolved and responsibility for the various navigations not in private hands was transferred to local interests. In the Coalisland area the Primate, Governors and *custodes rotulorum* of the lakeside counties and the proprietors of the working collieries were to act as trustees for the control and advancement of inland navigation between the collieries and the sea.

Transference of responsibility for the canals in east Tyrone proved no panacea for the navigations concerned and as early as February 1788,[26] parliamentary petitions requesting the construction of either an adequate water or waggon way from Drumglass to Coalisland reflect the absolute failure of Ducart's Canal and the acute anxiety of the colliery proprietors to have an effective link between their pits and the head of the newly-opened navigation at Coalisland.

Nor was it the extractive industry alone which was feeling the need for a continuous navigation providing cheap, regular and efficient transport between the collieries and the lough shores eastward. In May 1789 parliamentary assent was given to a bill[27] designed 'to promote the linen industry in the several counties bordering on Lough Neagh by making a communication between the said lough and the collieries at Drumglass in the county of Tyrone', while in April 1797[28] the linen drapers of Counties Armagh, Down, Londonderry, and Tyrone presented a petition asking for the thorough overhaul of the entire navigation as the bleachers had been compelled to import English coal at considerable expense due to the inability of the Tyrone collieries to export sufficient to meet their demands.

Though waterborne traffic moving downstream from Coalisland during the remainder of the century fell far short of the anticipations expressed during its protracted construction, by no means all the blame can be laid on the condition of the navigation alone. The collieries themselves were not proving nearly so productive as had been expected and while traffic from this source was increasing slowly it would still have been quite insufficient to allow the navigation to make a profit. Besides coal, however, a variety of general merchandise was being attracted to the new waterway and in 1786[29] William Harkness, a Coalisland bleacher, gave evidence as to the dependence of this branch of the linen industry on water transport. Large quantities of flaxseed, provisions, grain, timber, rock salt, fish, and coarse hardware were soon moving to and from Coalisland across Lough Neagh from the Lagan and Newry Navigations.[30]

There can be little doubt, however, that the failure of the collieries[31] to provide a substantial, regular downstream coal traffic to Newry goes a long way towards explaining the unsatisfactory condition into which the Coalisland Canal rapidly deteriorated after 1787. So long as exports of coal from the pits at Drumglass, Congo, Gortnaskea, Brackaville and Annagher remained insufficient to stimulate general interest and support for the new canal, so the condition of the navigation channel and works continued to decline, until by 1795 it was observed[32] that 'while in this county are some very fine collieries the want of a more perfect inland water carriage contracts the operation of the many benefits which the site of these collieries presents'.

By the beginning of the nineteenth century the entire navigation from the Blackwater to Coalisland was in a ruinous condition and with yet another transfer of control, this time a return to a central corporation, the Directors General of Inland Navigation, a thorough overhaul and repair of the 4½-mile waterway had become urgent. To estimate the condition of the navigation and to draw up a scheme of repair and improvement, the Directors General sent one of their engineers, Henry Walker, to Tyrone in the spring of 1801 to inspect the canal course and works and make a preliminary estimate of what needed to be done. On 4 May[33] he reported that repairs totalling £2,500 were urgently required and work was set in hand almost at once. However, following allegations of defective engineering on the inland section of the Newry Canal, which was also being largely reconstructed under Walker's supervision about the same time, he was dismissed from his post in September of the same year and John Brownrigg, another of the Directors' engineers, was invited[34] to complete the renewals and improvements already begun. Before proceeding, Brownrigg made a detailed report on the navigation, drawing attention to the dangerous, unsatisfactory nature of the greater part of the canal and its works and concluding that 'to enable the publick to derive a partial benefit from the navigation by next spring . . . a repair of walls, new or repaired gates and dredging of levels are the only points which can be attended to this season'.[35]

This adverse comment caused the Directors General to call upon the opinion of a third engineer, Daniel Monks, as to the best means of restoring the canal course, locks and lock-houses, quays and basins and his report on 1 December 1801,[36] throws further light on the condition of the navigation at this time:

> the foundation on which the first or bason lock is laid is very bad, being a fine kind of sand and the lower sill and wall adjoining has sunk so as to render the whole useless . . . this lock has been taken down and rebuilt . . . it is uncertain to know the quantity of sheeting or plank required to repair the bottom until three feet thick of mud is first removed from off it. . . .
>
> The foundations of the second or 'dam' lock are also laid on sand, yet the walls are apparently staunch and settled . . . the bottom is covered with mud two feet and a half thick . . . nothing remains of the lower gates but their decayed frames . . . I understand there are new ones providing for this and the other locks and that part of the timber has been stolen which, with seeing some lying up and down on the works, convinces me more of the necessity of having an established store. . . .
>
> . . . on entering on the third level, which is 248 perches long, there is a view of the most rough, uneven and unsettled piece of canal that can be imagined . . . here has been from the commencement a source for employment, a constant struggle between the labour of man and the efforts of nature.

Recommending the erection of a store at the basin, with an official storekeeper, to safeguard supplies of timber, cut stone, and machinery, and to prevent pilfering, Monks complains of inefficiency in the use of the ample supplies of cheap labour available, noting that with hundreds of labourers employed there are no more than twelve barrows and twenty planks.

All three reports to the Directors General reflect the poor condition into which the canal had degenerated since 1787 and determined efforts were made during the next ten years to improve the situation. Amongst the works carried out under the Directors General were a complete scouring and dredging of the Coalisland basin with the construction of adequate boundary walls, wharves and stores, the dredging and deepening of the whole navigation downstream of Coalisland to a depth sufficient to accommodate lighters drawing 4 ft 6 in., the reconstruction of the double lock under Benjamin Pemberton, the rebuilding of several lock walls and the sheeting and paving of lock floors, the replacement of several pairs of lock gates, the repair of three of the lock-houses, the staunching and puddling of large stretches of the banks in the lower sections towards the Blackwater, and a levelling and gravelling of the towpath.[37] Between 1801 and 1812 over £20,000 was spent,[38] yet when Wakefield visited the area in 1809,[39] like McEvoy before him,[40] he found the canal 'nearly choked up with weeds' with 'but little trade in the neighbourhood'. Ducart's Canal had already been abandoned and was soon being deliberately filled in and cultivated. In fact, Drumglass coal continued to be carted by road eastwards to Coalisland throughout the greater part of the nineteenth century, and no alternative, such as John Richardson's plan of July 1826 for a railway with two inclined planes, was ever adopted.

The passing of the Act of Union had indeed removed the basic motive for creating the line of inland navigation west of the Blackwater: the Irish Parliament's anxiety to have Tyrone coal conveyed to Dublin cheaply and in large quantities. With free trade between Ireland and England, English coal could be more readily supplied to the Dublin market at 30s per ton, a price sufficiently low to preclude Tyrone fuel which at this time had a pit-head price of 16s per ton, equivalent to 26s to 28s in Dublin.[41] Though slightly cheaper, the Irish coal was generally considered much inferior to English and Scottish imports,[42] and as the Dublin market came to favour cross-channel sources more and more, so the demand for Tyrone fuel declined and with it coal traffic on the newly-reconstructed navigation.

As the century advanced, however, canal traffic in a great variety of general merchandise showed a steady improvement: grain, tiles, fireclay, earthenware, coal, spades, and shovels on the outward run, and timber, slates, iron, salt, flaxseed, groceries, and lead on the inward. Coalisland, Moy and Maghery became nodal points in a network of collection and distribution, and between these towns, the Lough Neagh ports and the Lagan and Newry Navigations traded 70-ton lighters paying tolls at a number of points *en route*. The journey from Coalisland to Belfast, a distance of 61 miles took three days, from Coalisland to Newry, 39½ miles, two days.[43]

Taking the year 1836 for closer scrutiny, waterborne traffic moving to and from Coalisland was made up as shown overleaf.

Inwards

Type of traffic	Tons	Origin
Salt	300	Belfast and Newry
Square timber	1,160	Belfast and Newry
Planks and deals	400	Belfast and Newry
Staves (number 15,000)		Belfast and Newry
Herrings (barrels 150)		Liverpool
Sheet and lead ore	20	Liverpool
Slates	730	Wales
Bar and wrought iron	450	Newport, Bristol, Liverpool, Glasgow
Coal	nil	Occasional shipments, Belfast and Newry

Total   3,310

Outwards

Type of traffic	Tons	Destination
Flour	500	Belfast and Newry
Oats	1,400	Belfast and Newry
Potatoes	55	Belfast and Newry
Bricks and tiles	1,308	Belfast and Newry
Coal	718	Portadown

Total   3,981

Based largely on the local supplies of coal, surface clays and ease of communication by water, a number of industries developed in and around Coalisland, which gave the area an atmosphere more akin to that of rural areas of the English Midlands or of remote areas of Yorkshire and Lancashire than that of the north of Ireland at this time. Thus the manufacture of tiles, bricks, pottery and earthenware, fireclay goods, spades and shovels, sulphur, and sulphuric acid had all come to play a part in the life of the area by the 1830s and were contributing to waterborne traffic moving to and from Coalisland, both through the import of raw materials and the export of finished goods. In 1837 the town was described as 'a place of considerable trade, with thirty-five large lighters or barges, which frequently make coasting voyages to Dublin, and sometimes across the channel to Scotland . . . there are exports of coal, spades, shovels, firebricks, fireclay, crucibles, earthenware, linen cloth, wheat, oats and flour etc., while the imports are timber, deals, iron, salt, slates, glass.'[45]

In 1841 the Blackwater was linked to Upper Lough Erne by the 45¾ mile-long Ulster Canal from Charlemont to Benburb, Caledon, Middletown, Monaghan, and Clones and it was hoped that by building a navigable channel across the lake-studded surface of County Leitrim to join the Shannon near Drumshanbo at the southern end of Lough Allen a continuous line of inland water communication would be established across Ireland, comparable to the Grand and Royal Canals farther south. In 1860 this extension, the Ballinamore & Ballyconnell Canal, was indeed opened, but proved a complete failure.[46]

Hopes of establishing a flourishing traffic from Coalisland along the Ulster Canal itself, and to Belturbet, Enniskillen and Belleek, also remained largely unfulfilled, and though several manufacturers in the town did adopt the new line of water communication and develop a small but regular trade in fireclay goods, earthenware, agricultural hand tools, grain, and coal, the difficulty of direct intercommunication resulting from the narrow locks and shallower channel of the Ulster Canal did much to restrict and discourage through traffic to Upper and Lower Lough Erne.

A few years after the completion of the Ulster Canal a far-reaching scheme of waterway construction and improvement was begun on the Upper and Lower Bann, the Blackwater and around the shores of Lough Neagh, aimed at improving flood control throughout the entire basin and fostering inland navigation over a wide area extending north and south of Lough Neagh.[47] It was thought that Coleraine might well become a flourishing seaport, connected by a broad navigable artery to a large inland lake from which further lines of water communication led eastwards to Belfast, south-eastwards towards Newry and south-westwards along the Blackwater towards Coalisland, the Ulster Canal and Lough Erne. So situated, it could handle through traffic for much of north central Ulster and tap large areas in the Lough Neagh basin. The explosive rate of growth of Belfast during the period 1840–80, a rather later emergence than in comparable English industrial cities, linked with the development of a wide range of manufactures and the creation of a great port in the flat, shallow estuary of the Lagan, established that city as the focal point for the whole of the north, a development emphasized by the building of railways radiating from the city. Therefore traffic on the Lower Bann did not emerge on anything like the scale anticipated by the Irish Board of Public Works before or during the period of construction, though the improvements carried out on the channel and at the mouth of the Blackwater, and on the bridges at Charlemont and Verner's Bridge, and the erection of quays at Moy and Blackwatertown were developments eagerly welcomed by the merchants and traders of Coalisland who saw in them a means whereby the commercial and industrial prosperity of their town could be advanced further.

The extension of the railway from Portadown to Dungannon, reached in 1858 and Omagh, reached in 1861, had no immediate effect on waterborne traffic in the area, while the improvement and more widespread adoption of steam power in tugs and lake steamers speeded up the movement of lighters and barge trains on Lough Neagh, the Lagan and Newry Navigations, the Upper and Lower Bann, Blackwater and Lough Erne.[48] Indeed, under the Board of Public Works, to whom control of the canal had passed from the Directors General of Inland Navigation in 1831, a slow but fairly evenly-sustained increase in both import and export traffic did occur as the century progressed—from a total traffic of 8,200 tons in 1837 to 13,268 tons in 1844 to 17,787 tons in 1856 to 18,888 tons in 1866[49]—but with the rapid growth of railways encircling Lough Neagh, a process completed by the junction of the Great Northern and Northern Counties systems at Cookstown in 1879, the relative importance of water traffic experienced a fairly abrupt decline.

Indeed, in spite of this slow growth of canal traffic, and despite increases in tolls in 1833 and 1874, annual receipts during the remainder of the century never exceeded £350, a figure soon overtaken by rapidly-growing operational and maintenance expenditure. Between 1873 and 1878, for example, annual receipts averaged a mere £262, of which approximately £218 were from tolls and some £44 from rents, while expenditure came to nearly £400 in each year.[50] Thus, though the canal was being maintained in reasonable working order, providing a waterway link between 'Coal Island in Tyrone which is the beginning of a little "Black Country" in Ireland',[51] Newry, still a port of some importance following the partial clearance of the Carlingford bar between 1866 and 1868, and Belfast, whose remarkable expansion was creating an enormous demand for building sand and bricks, it was no longer realizing a working profit for the Public Works Commissioners. With a growing deficit on annual working, the failure of the Ulster Canal following its reopening in 1873 (see p. 110), and the rapidity of rail construction on both the Great Northern and Northern Counties systems during the 1870s and 1880s, the

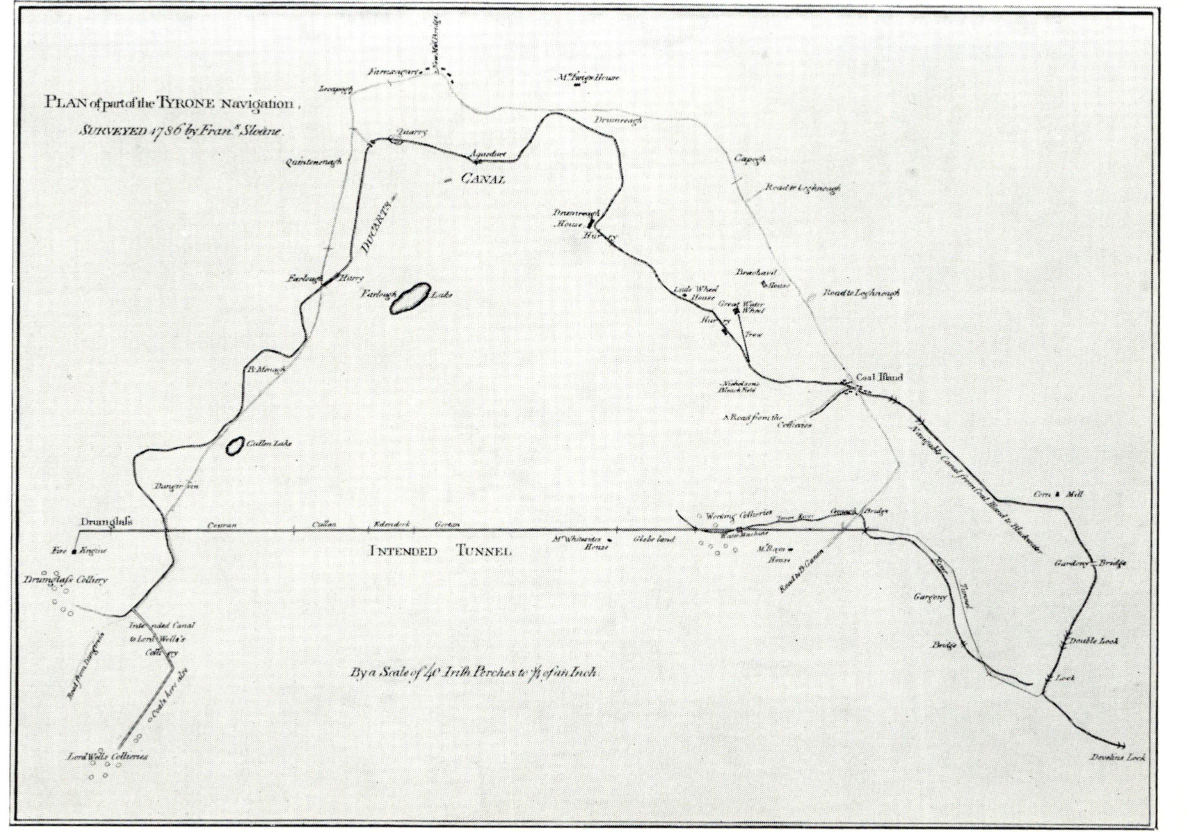

XI. Plan showing Richard Owen's proposed navigable tunnel to Drumglass

XII. Ducart's Canal: the inclined planes or 'dry hurrys' at (*above*) Drumreagh and (*below*) Farlough

XIII. The Strabane Canal *c.* 1910: (*above*) the Strabane basin from the south-east looking towards
Canal Quay; (*below*) ice-bound craft in the basin

XIV. Strabane Canal: (*above*) Crampsie's, the lower of the two locks on the canal; (*below*) the basin from the north-west, *c.* 1930

Government became anxious to rid itself of this small but continual drain on its finances and began to look around for some unsuspecting private company on which to foist not only the Tyrone Navigation but also the Ulster Canal, whose annual trading deficit had by now reached approximately £1,000. Eventually, in 1888, following a great deal of preliminary skirmishing, both canals were transferred from the Commissioners of Public Works to the Lagan Navigation Company (see p. 115).

Under the Belfast company, the canal bed was deepened by some three inches, giving a 5 ft draught and enabling barges carrying up to eighty tons to reach Coalisland. From 1892 onwards traffic on the navigation increased more rapidly than at any previous period, with numerous through cargoes moving directly to and from Belfast—sand, bricks, tiles, pottery, fireclay goods, agricultural products, and native timber on the outward run, grain, hardware, foreign timber, provisions and coal on the inward—and a trading deficit of £89 for the twelve months ending 31 March 1890 had become a profit of £355 by the end of the century, a figure representing a doubling of tonnage, from approximately 18,000 tons in 1890 to over 36,000 tons in 1900.[52] The improvement was well sustained through the early years of the present century, little hint being given of the rapid collapse soon to overtake all canal traffic in the north of Ireland. Indeed, there was talk of increasing the draught of the navigation to 5 ft 6 in. to facilitate through traffic from the Lagan Navigation and of improving both the water supply and the inland harbour at Coalisland, but none of these proposals had borne fruit before the outbreak of war in August 1914. After the war the rapid development of private and commercial road transport put paid to any schemes of canal extension or improvement.

Before moving into the final years of the long and complex history of canal construction and operation in east Tyrone, however, let us return for a moment to the pre-war era, select a year at random and analyse canal traffic moving to and from Coalisland to find out what changes had come about since the 1830s and 1840s. The years chosen are 1844 and 1914 and the figures are given overleaf.[53]

Imports

1844	*Tons*	1914	*Tons*
Merchant goods	2,167	Indian corn	14,000
Coal	495	Artificial manures	1,000
Salt	550	Coal	5,500
Slates	500	Bran, flour and	
Foreign timber	533	provisions	2,400
Native timber	250	Timber	650
Wheat	290		
Total	4,785	Total	23,550

Exports

1844	*Tons*	1914	*Tons*
Wheat	425	Potatoes	1,200
Oats	2,150	Grass seed	500
Coal	1,325	Hay	400
Bricks and tiles	2,985	Sand	11,000
Pottery clay	400		
Home timber	110		
Flour	90		
Meal	90		
Sand	100		
Flax	723		
Stones	85		
Total	8,483	Total	13,100

Two important changes are evident; a decrease in the range of
goods handled, a reflection of the usurpation of the waterway by
the railway as carrier for such items as merchant goods and provisions, where speed of delivery was of greater importance than
cheapness of transit, and a complete reversal of the ratio of imports to exports. Further, an annual coal export of some 1,500
tons in the 1840s had been replaced by a much larger import by
1914, a reflection both of the cessation of local coal working and
of a certain geographical momentum in the industrial life of
Coalisland and its environs; in 1844 over 2,500 tons of wheat and
oats had been exported from Coalisland by water, by 1914 this
traffic had disappeared and been replaced by a large import of
bran, flour and maize; the amount of building sand exported in
1844 had been increased a hundredfold by 1914 to meet the enormous demands that had accompanied the rapid expansion of

Belfast, while the import of over a thousand tons of artificial manures in the latter year reflects the introduction of scientific farming methods.

Perhaps most significant of all is the huge development of traffic in Indian corn or maize, which accounted for well over half the import tonnage in 1914 and exceeded the total traffic, both import and export, for 1844 by almost a thousand tons. At this time maize was being brought from Belfast in 80-ton lighters which had been towed alongside the grain vessels at Belfast docks and filled by direct suction funnel loading. These barges, belonging mainly to the freight-carrying Inland Navigation Company, were then hauled up the Lagan Navigation by horses, the journey from Belfast to Ellis' Gut costing approximately 20s with tollage dues at the rate of 10d per ton. From the Gut a steam tug, owned and operated by the Lagan Navigation Company, took the barges across Lough Neagh and up the Blackwater, often in short barge trains. This cost 14s per lighter on the outward run, fully laden, and 10s on the return journey, empty or only partly laden. From the canal mouth near Verner's Bridge up to the canal basin horse towing cost about 5s per barge with tolls of 5d per ton. Thus, 80 tons of grain could be conveyed from Belfast to Coalisland for some £6 or roughly 1s 6d per ton, a figure comparing very favourably with that current on the competing Great Northern line from Dungannon.[54] Moreover, the corn mills at Coalisland were alongside or closely adjacent to the canal basin and at some distance from the railway, which had the further disadvantage of a somewhat circuitous route from Dungannon. After crushing, maize meal was distributed by 'road motor waggon', as far afield as Pomeroy, Ballygawley, Castlecaulfield, Moy, and Coagh at prices with which Belfast millers, using rail transport, were unable to compete. This meal formed an excellent fattening for pigs and the importance of pig rearing in the area around Cookstown owes much to the ease and cheapness with which grain could formerly be imported in large quantities to the mills at Coalisland.[55]

From 1 July 1917 until the establishment of a separate Government in the north of Ireland, the Lagan, Ulster, and Coalisland Canals were all placed under the direct control of the British Government. With return to private control, a grant of some £19,000 for improvements and works of reconstruction on both the Lagan and Tyrone Navigations, and the re-emergence of some semblance of order and security with the establishment of a new

Government in the north, the immediate post-war years seemed to hold fresh promise for canal transport in and around the Lough Neagh basin. There was a steady increase in exports of building sand and imports of grain and coal during the 1920s, until by 1930 and 1931 traffic had reached an all-time record of some 57,000 tons in each year. Toll receipts, too, increased rapidly, from £811 in 1922 to £1,634 in 1931, the latter representing a trading profit of £650.[56] As is usual, however, with trends occurring out of harmony with broader changes in the structure of an area's industry, commerce or internal transport system, the canal boom burnt itself out within two or three years and the resulting collapse of the navigation as a commercial waterway was thereby thrown into marked relief.

The reasons for this collapse are not difficult to find. Firstly, the development and improvement of the internal combustion engine as a direct result of the demands of the wartime emergency, coupled with the wholesale disposal of surplus road vehicles by the Government after the war led to the rapid growth of commercial road transport, at first in the hands of private hauliers, then by road services provided by private concerns and also the main railway companies, and, later, by a nationalized transport board. This, coupled with the slowness of canal transport in relation to the demands of modern trade and industry, led to a transference of traffic to road and rail. As early as 1924 the canal was regarded as unsuitable to carry large quantities of coal from Sir Samuel Kelly's new collieries at Annagher near Coalisland and railway sidings were constructed to handle the anticipated coal export to Belfast.[57]

By the mid-1930s competition from road lorry and rail waggon had begun to exert a final stranglehold on canal traffic, especially that moving downstream from Coalisland. The greater mobility and speed of road carriage from the numerous sand pits in the area and around the lough shores to eastward was soon making severe inroads into the substantial downstream sand traffic and between 1932 and 1937 the tonnage exported by water fell from 19,575 to 2,434. Furthermore, with the failure of the Annagher pits and the increasing fuel demands of the expanding fireclay works alongside the canal basin, a growing coal import could have been expected. Yet the import of coal by water also declined markedly during the decade, from 15,480 tons in 1932 to 8,701 tons in 1939. Finally, upstream grain traffic, so long one of the chief sources of income, also decreased, though not so abruptly.

By the eve of the outbreak of war the annual trading profit was down to less than £50.[58]

In effect this rapid fall in tonnage and revenue marked the end of the long history of the Tyrone Navigation. Traffic during World War II was insignificant, finally ending in 1946. The Coalisland Canal was abandoned in April 1954, in which year responsibility for it as a drainage channel passed to the Ministry of Commerce in the Government of Northern Ireland. In 1962 it passed to the Ministry of Finance.

In keeping with a general governmental policy of closure of inland waterways throughout the province, determined largely by economic necessity, the long era of canal transport to and from the mining and manufacturing area of east Tyrone has come to an end. The canals have given way to road transport, a medium in many ways less suited to the type of goods for which transport is still required, but fulfilling certain essentials in the modern commercial complex.

CHAPTER V

The Strabane and Broharris Canals

◆◆

THE Strabane Canal, constructed between 1791 and 1796 by the Marquis of Abercorn at a total cost of £11,858,[1] extended from the town of Strabane for a distance of just over four miles to a junction with the Foyle, on the extreme northern edge of the Parish of Leckpatrick. The decision having been taken to cut a navigation channel northwards from the town to the broad tidal waters of the Foyle, negotiations were entered into by the Marquis's agents with various smallholders in the townlands of Dysert, Greenlaw, and Ballydonaghy, along the line, and the requisite lands purchased.[2] Richard Owen, engineer for the extension of the Lagan Navigation from Lisburn to Lough Neagh, was consulted as to the line proposed and construction began late in 1791 under John Whally of Coleraine, County Londonderry. By October of the following year the channel had been extended northwards through the Parish of Leckpatrick and across the Ballymagorry river. It was not until 1795, however, that the locks were completed, entry of the main feeder achieved and junction with the Foyle accomplished just south of the entry of the Burndennet river, opposite Porthall in County Donegal.[3] The canal was opened to public traffic on 21 March 1796 amid confident anticipations of immediate commercial success:

> on Wednesday last the Strabane Canal was opened for the reception of vessels. When we reflect the general use that will arise from such a navigation, in the heart of a popular country, extending upwards of four miles from the town of Strabane where it joins the river Foyle, and the very great breadth and depth of it, being the largest in Europe, we cannot avoid praising the noble and liberal patron of such an undertaking.[4]

The scenes attending the official opening of the canal bear a close resemblance to those which occurred on the Lagan and Newry Navigations earlier in the century:

> early in the morning all the boats on the river assembled at the mouth of the canal in order to try their dexterity on the still-water navigation and to see which would have the honour of first arriving at Strabane: the inhabitants of all the country around assembled on the occasion, each to assist his friend, when to his honour who first caused the drawbridge of Derry to be opened, Mr Fleming's boat, Captain Quigley, was the first that arrived at the quay. In the evening a number of respectable inhabitants met and dined at the Abercorn's Arms and drank many loyal and other toasts suitable to this day: the populace was regaled with ale and bonfires and illuminations and other demonstrations of joy closed the night.[5]

The canal had a total length of 4 miles and 6 chains. Its main source of water supply was a stream which entered on the eastern side at a point about a mile and a half from the Foyle, and had its source in Moorlough, on the slopes of Knockavoe on the western edge of the Sperrins. There were two large locks in its lower reaches, one being 108 ft long and 23 ft wide with a depth of 7 ft on the sill, the other some 117 ft long, 24 ft wide, with a depth of 6 ft 6 in. on the sill. The first of these locks, in the townland of Greenlaw, divided the canal into two pounds of unequal length, while by the latter the canal was led down to the tidal reaches of the Foyle some ten miles above the quays at Londonderry. Both were designed to take ocean-going schooners of some 300 tons burden and it would seem that by the construction of such a waterway, the Marquis hoped to stimulate the industrial and commercial life of the town of Strabane and its environs, most of which lay within the boundaries of his estates.

The Strabane Canal followed hard on the heels of the Coalisland Canal, completed in 1787, and the headward extension of the Lagan Navigation from Sprucefield, above Lisburn, to Lough Neagh, completed at the end of 1793. Both before and after its construction the idea of linking Lough Foyle and Lough Swilly had been discussed (see p. 149) but nothing ever came of this idea.

Abercorn agents charged a standard rate of 2s per ton on the traffic in coal, timber, hardware, and foodstuffs which soon developed on the Strabane Canal. Traffic was at first mainly between Londonderry and Strabane and despite dissatisfaction amongst local merchants and traders, who regarded the 2s charge as both excessive and unreasonable, the early years of the nineteenth century saw the growth of a substantial barge traffic, mainly in

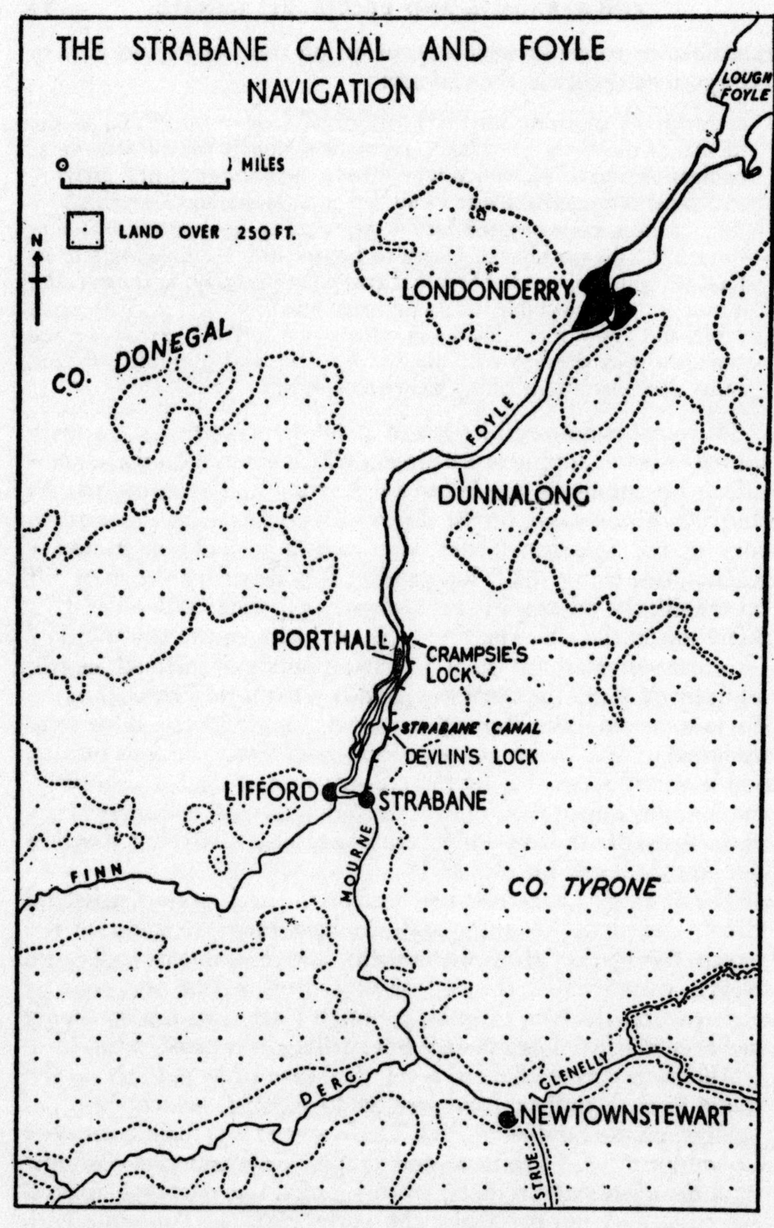

11. Map of the Strabane Canal, etc.

agricultural produce, from Strabane to the point of export at Londonderry, which was now handling a considerable mercantile traffic.[6] Coasting craft often moved upstream as far as Dunnalong, five miles above the town, where they could lie in nine or ten feet of water, before entering the canal. Lighters were towed by a small steamer as far as the entrance, and thence by horses to Strabane.

In 1820 the canal was leased to a number of interested parties in Strabane and district and at first its operation under this new body continued to be attended by a fair measure of success. For example, in 1836 canal lighters, each manned by two men, made 583 voyages between Strabane and Londonderry, and 10,535 tons of merchandise, mainly grain, were handled.[7] Strabane had by this time a population of over 4,700, and had become a flourishing market town at the junction of a number of fertile lowland corridors. It had the largest grain market in Tyrone and, in addition, an extensive trade in provisions, especially beef, pork, and butter. Its exports included wheat, oats, barley, flax, pork, beef, butter, eggs, poultry, beer, and ale: its chief imports were timber, iron, groceries, and coal.

There can be little doubt that the Strabane Canal was of considerable importance to the industrial and commercial life of the town in the pre-railway era—'its banks lined with large ranges of warehouses and stores for grain, with wharves and commodious quays well adapted to the carrying on of an extensive trade'.[8]

Soon, however, a variety of factors began to operate against the navigation and its controlling company. In April 1847, a single track standard-gauge railway line was completed from Londonderry to Strabane, crossing the low-lying, poorly-drained flood plain of the Foyle by means of numerous high embankments and bridges. This line, which had been surveyed by Robert Stephenson, had by 1852 reached as far inland as Omagh while in August 1854, Enniskillen was linked to Londonderry through Fintona, Omagh and Strabane. The Londonderry & Enniskillen Railway undoubtedly bestowed considerable benefits on the towns of the Foyle valley which it served and indirectly on the arable farmlands of the valleys and pastoral lands of the flanking uplands to east and west. Omagh and Strabane, in particular, emerged as major market centres from which goods and livestock could be conveyed northwards to the port of Londonderry. In 1861, when Omagh was reached from the south-east across the watershed separating the Foyle and Bann catchment areas, rail links were established between Londonderry, Belfast, and Dublin.

When the initial section of railway between Londonderry and Strabane was first opened, only passenger trains were run from each terminal daily but soon goods and livestock services were introduced and with the extension of the system to Omagh and Enniskillen, and its later connection with Belfast and Dublin, rail traffic expanded fairly rapidly. This in turn affected traffic on the canal and the controlling company soon found itself in serious financial difficulty.[9] In 1860 the Londonderry & Enniskillen Railway Company's lines were leased to the Irish North Western Railway Company which had constructed the 70-mile section from Dundalk to Enniskillen, making through services from Londonderry to Dublin easier and adding greatly to the commercial potential of the northern port. So abrupt was the movement away from the canal, indeed, that for the years 1864 to 1867 inclusive, the canal company obtained an average annual net revenue of just over £200 and the Duke of Abercorn, from whom the waterway was leased, a mere £450 per annum as a proportional levy in lieu of a fixed rent.[10]

Already, however, the canal had changed hands, for on 1 July 1860 the canal company, which had been operating the navigation on lease since 1820, was compelled to liquidate its few remaining assets and retire from the scene. A new company, the Strabane Steam Navigation Company, sometimes called the Strabane Steamboat Company, took over the lease from the Duke for the unexpired part of a term of twenty-five years, with the option of renewal for a further twenty-five years. Traffic on the canal continued at about 20,000 tons a year during the remainder of the century with only minor oscillations, realizing a gross annual revenue for the Steamboat Company of between £2,000 and £3,000. Annual net revenue, however, never exceeded £300 after the rent to the Duke, cost of management and maintenance and the carrying business had been met, and it came as no great surprise when the Strabane Steamboat Company also went into voluntary liquidation. On 28 April 1890, the Strabane Canal Company was incorporated and took over the navigation on lease from the Duke of Abercorn. Their lease was to be for an initial period of thirty-one years at an annual rent to the Duke of £300. The objects for which the new company had been formed were 'the conveyance of passengers and goods in steamboats, barges, lighters and boats, between the town of Strabane and the City of Londonderry'.[11]

By now, however, the general condition of the canal had deteriorated. Moreover, there was considerable friction between the new company and the Duke on the one hand, and a large number of Strabane traders on the other. Under the Railway & Canal Traffic Act of 1888, it had become obligatory for every public company to lodge lists of rates and charges with the Board of Trade for inspection and approval. The Strabane Canal Company contended that it was a private company and therefore lay outside the terms of the Act and the Duke of Abercorn, as owner of the navigation, lodged a petition against the Provisional Order of 1890 in the House of Lords. This led to an official inquiry into the affairs and by-laws of the company, which aroused considerable local feeling and led to a detailed inspection of the waterway by a representative of the Board of Trade, Lieutenant-Colonel Addison.

It had been alleged by traders that the canal company had failed to maintain the canal and works in a satisfactory condition, as it was bound to do, so that navigation on it had become increasingly hazardous. On this complaint Addison concluded that

> while the east bank of the waterway is undoubtedly frail and liable to be a cause of much anxiety in the future if it is not strengthened, and although there are grounds for complaint as to the depth of water along various sections and also as to the condition of the canal gates, it cannot be fairly said that the canal is in an unnavigable state at the present time.[12]

The second main point of grievance had been over the tolls. The canal company had been anxious to have a standard toll of 2s on all merchandise, whether carried in bye-traders' vessels or in their own, but this was disallowed and a tonnage rate of 6d was fixed by the Board of Trade on what was deemed by them to be a public company.

As the present century opened, therefore, memories of the public inquiry and the bitter animosities it had occasioned were still fresh in the minds of merchants and traders in the area; it was an atmosphere in which genuine co-operation with the canal company was most unlikely. Moreover, since the Government inquiry of 1898 had not specifically stated that the condition of the canal channel, locks, and terminal facilities was inadequate, nothing was done subsequently to improve them; as a result commercial navigation was further discouraged.[13] By 1910, for example, large stretches of the canal were less than 2 ft deep, mainly because of leaky locks and banks, lighters often having to travel half-loaded, and more than two miles of the east bank required to be strengthened and made watertight. In the Foyle, half a mile from the canal entrance, lack of adequate and regular dredging by the Londonderry authorities who were responsible had resulted in the growth of shoals and sandbanks on which small vessels often went aground, even during ordinary tidal conditions. In the early part of the nineteenth century it had been normal for coasting craft to use the canal, and for livestock and provisions to be shipped direct from Strabane to Glasgow. Indeed, it had then been possible for steamers to reach Castlefinn, while the Foyle was used regularly for the delivery of coal to Lifford. However, under the Steamboat Company it had become the practice to transfer Strabane traffic to lighters at Londonderry, to be towed to the canal. By the end of the century, though harbour dues were still being paid at Londonderry upon Strabane traffic, it had long been quite impossible to bring large vessels upstream to the mouth of the canal.

Indeed, after the construction of the Carlisle Bridge at Londonderry in the 1860s and, later, of the narrow-gauge railway[14] bridge across the canal itself a short distance north of the basin at Strabane, which gave only 14 ft or 15 ft headroom and reduced the canal width to 18 to 20 ft, it became impossible for sea-going vessels to reach the town. Large craft, usually moving to and from Londonderry were compelled to lower their masts or funnels when passing under this bridge. At two other points on the canal wooden swing bridges could be moved aside to allow the passage of vessels.

Waterborne traffic to and from Strabane at the turn of the century included Indian corn, oats, flour, coal, pork, potatoes, flax, eggs, butter, poultry, hardware, and timber. Particularly important was the traffic provided by Smyth's grain mill, situated alongside the canal basin, in some years amounting to half the total of 20,000 tons. A jetty was built into the canal behind the gas works to facilitate the unloading of coal lighters from Londonderry, while alongside the canal basin were situated two saw mills, a brewery, a tannery and several repair sheds and docks connected with the servicing of the company's vessels. At this time an average of two or three boats a week reached Strabane, each containing between 40 and 65 tons.[15]

Despite its largely unsatisfactory condition, the canal performed a further service to the town of Strabane and the various districts through which it ran. Prior to the construction of the narrow-gauge line from Strabane to Victoria Road, Londonderry, by the

Donegal Railway Company, opened in August 1900, the only competition encountered by the Great Northern Railway between Strabane and Londonderry had been that of water transport on the Foyle. Merchants in Londonderry and Strabane had hoped that by helping the construction of a second railway between the two points, healthy competition would be engendered and this in turn would lead to a reduction in tariffs on the standard-gauge system.

In fact, it seems that in the years immediately following 1900, a mutual arrangement was arrived at between the two railway companies which emphasized the important role played by the Strabane Canal and the Foyle in keeping down rates on both the parallel railways. In 1910 there was a difference of about 6d per ton in favour of water transport between these two points, 2s (including the 6d toll laid down by the Board of Trade) for bread-stuffs, feeding stuffs and manures and 4s 6d, excluding porterage at Londonderry and cartage at Strabane, for all other merchandise. Moreover, the canal basin at Strabane lay considerably nearer the centre of the town than did the railway station.

With the death of the principal shareholder in the canal com-pany (James McFarland, the Londonderry contractor), disagree-ment amongst the other members, consequent lack of managerial authority and a further deterioration in the condition of the navi-gation resulted in a decrease in both traffic and net revenue. Eventually, in 1912, the whole of the navigation passed to the Strabane & Foyle Navigation Company Limited. This time the canal was sold outright by the Duke, though he remained a tenant for life under the conditions of a former marriage settlement. The new company, the controlling interest in which was at first held by William Smyth, the Strabane miller, had a capital of £3,000 in £1 shares. Attempts were made to improve the draught of the naviga-tion and a steam tug, the *Shamrock*, was acquired to tow barge trains up to the canal basin at Strabane but, despite this, traffic diminished fairly rapidly during the succeeding years and finally disappeared in the early 1930s. From 1944 efforts were made to have the canal abandoned and eventually, in December 1962, after much negotiation, the section from the basin at Strabane to the site of the former swing bridge in the townland of Dysert was officially closed. Technically the rest of the canal to the Foyle remains open.

The canal bed and all its tow paths, locks, docks, wharves and dwelling houses still belong to the Strabane & Foyle Navigation

Company. The canal basin at Strabane has been entirely filled in, though the former quayside sheds and warehouses, cranes and bollards are still in place around its three sides. Downwards on the west bank can be seen the remains of the former dry dock and basin where the *Shamrock* and her charges were taken for periodical overhaul and repair while farther north, on the same bank, the timbers of the old jetty at the rear of the gas works still project into the broad, overgrown ditch that is now the former canal.

The low-level railway bridge that formerly carried the narrow-gauge line from Strabane to Londonderry has been demolished, but its massive stone supports still remain in position. The two wooden swing bridges, formerly operated by resident employees of the company, have both been removed and replaced by ramps carrying rough roadways over the canal bed. The masonry of the two large locks is still sound, but the condition of the massive lock gates is less so, though much better than that of those remaining on the Newry Navigation or the Ulster Canal.

The condition of the canal bed varies considerably; there is much more water in the lower reaches, below the entry of the main feeder, than in the upper section towards Strabane, which is merely a broad, marshy strip between raised banks, in places entirely filled in by recently-dredged material from the rivers Mourne, Finn and Foyle.

The shortest of all the inland navigations in the province, the influence of the Strabane Canal was much more restricted than, say, that of the Lagan or Newry Navigations. Though only slightly shorter than the Coalisland Canal the traffic which it carried was, in general, much lighter. In east Tyrone an industrial enclave based on local deposits of surface clay and coal provided downstream traffic during the first half of the nineteenth century, and though the export of coal declined markedly as the century progressed, it was replaced by such goods as freshwater sand, gravel, bricks, and an increase in the output of fireclay goods.

With the Strabane Canal, which had been designed with the intention of linking the main market town on the Marquis of Abercorn's Irish estates not only with the main seaport of the area, Londonderry, but also with cross-channel ports, especially Glasgow, agricultural produce, livestock and provisions always provided the main outward traffic, coal, timber, hardware, grain, and foodstuffs the main items being imported through Londonderry from much farther afield. From the size of the locks and the breadth of the canal course it is obvious that the original intention was to establish a regular, direct service by schooner between Strabane and cross-channel ports, but though such traffic did develop to a limited extent in the early years, with deterioration in the condition of the navigation channel both in the Foyle itself and in its headward extension to Strabane, it soon became impossible for ocean-going vessels to negotiate the canal or even the shallow, tortuous course of the Foyle above Dunnalong. After the building of the Carlisle Bridge at Londonderry in the 1860s, transhipment to lighters at Londonderry became the rule.

Though it passed through a number of hands, the canal never yielded a net revenue of more than £300 per annum, and from some 20,000 tons carried annually at the beginning of the century, traffic declined to such an extent that the income from grazing rights along the banks soon exceeded that from tolls and haulage. Whilst it contributed to the growth of Strabane as an important market town in the early years of the nineteenth century and, later, acted as a temporary curb on railway rates, in common with most other inland navigations in the province the canal soon succumbed to the competition provided by parallel lines of road and railway.

Broharris Canal

Below Londonderry there was a short cut, some two miles in length, near Ballykelly in County Londonderry, from Ballymacran Point on the southern shore of Lough Foyle southwards towards Limavady. Generally known as the Broharris Canal, this served both as a drainage channel and a navigation cut. It was constructed in the 1820s at a cost of some £4,500 and for a time handled a certain amount of traffic in heavy goods, raw materials, and bulky foodstuffs moving down the Foyle from the quays at Londonderry.

Its main importance, however, was as a navigable channel along which large quantities of shellfish and kelp were brought in from the banks exposed at low tide along the shallow, shelving coastline stretching westward from Magilligan Point. These were used extensively as fertilizer on the sandy soils on the flat land around the foot of Benevenagh. It was probably the moderate success of this cut which prompted the inhabitants of Limavady to press for a separate canal right up to the town from Lough Foyle at about the same time (see p. 151).

No mention of the canal is made in the report of the Irish Railway Commissioners in 1838, nor does it receive more than a passing mention in any of the Government reports of the second half of the nineteenth century. It is not listed in the final report of the Royal Commission.

The Strabane Fleet

(Original version, as taken down from Mr Thomas Lowry, Main Street, Strabane.)

> Come all you jolly seamen bold
>   That plough the raging main,
> Give an ear unto my tragedy
>   And I'll relate the same.
> Our 'Shamrock' slowly moved off,
>   And I in her did go;
> That very night at six o'clock
>   The stormy winds did blow.
>
> Her steaming works remained untouched
>   For two long hours or more,
> She logged and heaved most dangerously,
>   Not very far from shore.
> Her cabin windows were all broke
>   There scarcely were left one,
> When the Mate cries to the Captain
>   'Sir, we'll never reach Strabane.'
>
> When we came to the New-bridge,
>   No danger did we fear;
> Our Captain he stood on the deck
>   And told me for to steer.
> 'Oh, it's steer your helm and port, my boy,
>   With your bows towards the lan'
> For I think we'll have rough weather
>   Before we reach Strabane.'
>
> The raging seas rolled mountains high,
>   No mercy from the wave;
> We expected every minute
>   That we'd find a watery grave.
> The second shock the 'Shamrock' got
>   All hands were bound to cry:
> 'May the Lord have mercy on our souls
>   For near Prehen we lie.'

XV. Ulster Canal: (*above*) bridge and lock house at Milltown; (*below*) Charlemont lock (No. 1) at its junction with the Blackwater

XVI. Ulster Canal: (*above*) former canal stores at Monaghan town; (*below*) siphon under the canal at Clones

XVII. Upper Bann: (*above*) lighters and steam tugs at Portadown bridge, *c.* 1925; (*below*) warehouses in Castle Street, Portadown

XVIII. Upper Bann: (*above*) Newport Trench, once a busy port on Lough Neagh; (*below*) Kinnego Gut near Lurgan, a former harbour on the Upper Bann Navigation

We took our way to Carrigans,
  No danger did we fear,
But looking towards St Johnston
  No lighthouse did appear.
The raging seas rolled mountains high,
  And the wind was blowing strong,
When the Mate cries to the Captain:
  'Sir! Oh, yonder's Dunnalong!'

Oh! it's Dunnalong that seaport town,
  If we were landed there,
We would have the best of harbour,
  And no danger might we fear.
We'll steer our course for Dunnalong,
For I think it's our best plan,
Or else the 'Shamrock' might be lost—
  The fleet bound for Strabane.

The wind it changed to the North-West
  And dreadful was the night,
We looked out towards Porthall,
  But we could see no light.
It's 'count your men' the Captain cried,
  'For I think we've lost McShane'
It was a dreadful passage
  In the fleet bound for Strabane!

The crew being hearty all the way,
  They sang an old sea song:
'O, Molly, I love your daughter—
  I love no other one.'
The wind it changed to the North-West,
  And then came on a squall;
And all the grub we had on board
  Was a bottle of castor oil.

Thank God we landed in Strabane,
  No danger do we fear;
We'll drink a health to seamen bold
  and brave, while we lie here.
When looking over Derry Bridge
  There is nothing half so gran'
As to view the fleet that sails the deep,
  From Derry to Strabane.

G

CHAPTER VI

The Ulster Canal

THE history of this navigation spans little more than a century, from 1815 to 1931, and is largely a chronicle of insuperable practical difficulties and piecemeal organization. Though extensive, having a total length of some forty-six miles, it never enjoyed the success which attended the Lagan, Newry or Coalisland Navigations. Indeed, from the outset its efficient operation proved impracticable, largely because of fundamental errors in construction and, though forged as a link in what was hoped would become an important arterial waterway across Ireland, unforeseen natural difficulties and lack of administrative co-operation gave it little chance of success.

The navigation extended from Charlemont on the Blackwater to Wattle Bridge on the river Finn, south-east of Upper Lough Erne, and was constructed as an important section of a great composite waterway which was to be formed across Ireland, from Belfast to Limerick, by the cutting of a further canal across County Leitrim, from Upper Lough Erne to the Shannon. Once this was achieved a continuous navigational link would be created connecting Limerick, Athlone, and Carrick-on-Shannon with Coleraine, Belfast, and Newry, a link which would both supplement and compete with the Grand and Royal Canals farther south, both of which were already in existence and enjoying considerable success. Conceived as an important 'link navigation' rather than as a major waterway in its own right, it was imperative that to succeed it should be in a position to handle a great variety of through traffic passing from the Lough Neagh basin south-westwards towards the Erne lowlands and headwaters of the Shannon. It is therefore somewhat difficult to understand why those in charge of its construction should have immediately rendered the bulk of such traffic impossible by erecting locks of insufficient width or depth to handle the normal lighter from the

Lagan and Newry Navigations, and Lough Neagh. From a total expenditure on construction works of over a quarter of a million pounds, those in control of the waterway during its relatively brief life never recouped more than £500 per annum, a sum falling considerably short of that often required by normal maintenance and regular operational expenditure.

A proposal had been made as far back as 1778 for a canal from Murray's Quay at Ballyshannon upstream through Ballinacarrick, Cherrymont and Cumlin, to cross the river Erne by a two-arched aqueduct not far from Belleek and, passing through that village, to connect with Lower Lough Erne above the falls. The estimate for this scheme was some £32,000, with a further £8,000 for a navigation channel in Upper and Lower Lough Erne to connect the proposed canal with Enniskillen, Belturbet and Ballyconnell. 'Its future extension would be from the Upper Lake to Ballyconnell and by the Woodford River to Ballymore, Lough Scurr and to Communicate with the River Shannon at Leitrim. It would then open a passage from the northern to the western ocean.'[1] In 1783 the Irish Parliament made a grant towards the cost, and during the next ten years a section between Belleek and Ballyshannon was cut under the superintendence of Richard Evans, engineer of the Royal Canal, and a lock built at Belleek. Work was abandoned for lack of funds in 1794.

Although in 1801 the Directors General of Inland Navigation asked Evans for a revised estimate, no action followed.

During 1814 the idea of a link between Lough Neagh and Lough Erne probably took shape in the minds of the Directors General as a measure by which unemployment might be alleviated in the area and the lot of the Irish peasant improved over much of north central Ireland. On 1 November of that year John Killaly was ordered to make a survey of the area between the Blackwater and Lough Erne, and on 2 February 1815, he laid his report, with an accompanying map, section and estimate, before the Directors in Dublin.[2] Killaly estimated that a canal some 35½ miles long with six ascending and sixteen descending locks, from Wattle Bridge on the river Finn, south-east of Upper Lough Erne, to Charlemont on the Blackwater, would cost some £223,000. He also envisaged a branch to near Armagh.

'Not being limited by the Board's order to any particular size for the proposed canal', he says, 'I have calculated on one of similar dimensions to the Royal Canal extension, as being suited to the passage of boats capable of navigating the lakes.' Therefore presumably he provided for locks measuring about 76 ft by 14 ft. The narrowest lock on the Royal, however, was 13 ft 3¾ in. and we see, therefore, that Killaly had already departed from the ruling width of 14 ft 10 in. which prevailed on the Lagan, Newry and Tyrone Navigations. It was a mistake by the Board and by himself that was to have, in its later forms, disastrous consequences.

His evaluation of the engineering work which would be involved, and of the probable benefits of such a canal, are not without interest:

> the only waters which can be had for the summit level, some seventy feet above Lough Erne and one hundred and sixty feet above the Blackwater, are the heads of the River Finn, which runs through Wattle Bridge and the streams which flow from Drumaconner, Scotstown and Tedavnet, which I have strong hopes would be found a sufficient supply, particularly as I proposed sinking deep through the summit level, preparing it for eight feet depth of water . . . the line is free from any material difficulty until it reaches the falls of Benburb, where the Blackwater runs with great rapidity for about one hundred and twenty perches through a deep and rocky ravine . . . for this distance the line must unavoidably keep close to the river and the breadth of the canal be contracted so as merely to admit the passage of one boat, but as five locks will be required near this, I do not much regret the excavation through rock . . . on most other parts of the line the soil is of so soft a nature as to render the earthwork very easy of execution. . . . The country through which the proposed navigation would pass is in general very fertile in grain, particularly barley and oats; it also affords excellent pasturage, much butter, and large quantities of live cattle and pork are sent from here to Belfast and Newry. Limestone is abundant but the bogs being small and mostly cut out, occasions a great scarcity of fuel on many parts of the line.
>
> Great benefits would result to the province of Ulster from the execution of this canal by its affording a cheap mode of transporting the redundant produce of the Counties of Cavan, Monaghan, Fermanagh, Tyrone, etc., to the markets of Belfast, Newry and Armagh, which towns, besides participating in those benefits, would thereby find a ready vent for timber, iron, salt, flaxseed, groceries, etc.
>
> The canal would also be peculiarly inviting for the establishment of passage boats and highly advantageous in the conveyance of fuel from Lough Neagh and the towns I have mentioned to many parts of Monaghan, Armagh and Tyrone, not only for house use but also for the burning of lime, the farmers being at present, in a great degree precluded from liming their lands for want of turf or coal . . . the line between Lough Erne and Lough Neagh would in many

other respects be extremely desirable, particularly if a canal from Belleek to Ballyshannon were executed, as by those means a safe, internal navigation would be opened between the northern parts on the eastern and western sides of the kingdom, thereby avoiding the delay and danger attendant on a sea voyage between those places ... by the execution of those canals a door would be opened to the English market from the north-western parts of Ireland, agriculture would be encouraged, the comforts of the poor increased and a check put to the spirit for emigration, at present so prevalent in that part of the kingdom.

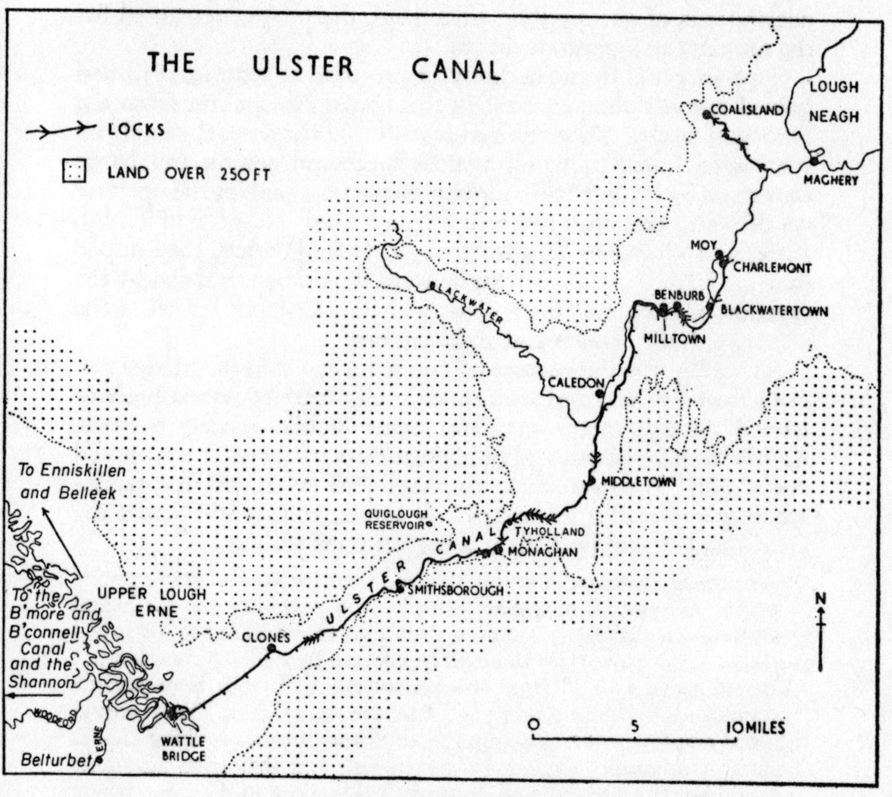

12. Map of the Ulster Canal

However, despite the fact that at a town meeting held at Monaghan in February 1817 the landed proprietors through whose properties the new canal would pass, together with a number of prominent Belfast merchants, offered to provide two-thirds of the estimated cost of construction, no attempts were made by the Directors General to implement Killaly's suggestions. Whether this was because they did not relish the thought of embarking on a new cut of considerable size, involving heavy expenditure, while there remained so much to be done by way of repairs and improvements on canals which had been in existence for some considerable time or whether the whole concept of linking east and west coasts had temporarily gone out of favour is not clear but despite numerous public meetings, memorials, petitions, and declarations of great public advantage, the project remained for the moment an ambitious dream.

Support came from the body of proprietors who in 1810 had gained a controlling interest in the Lagan Navigation from the Donegall family. They realized that the shortness of the hauls on the Lagan Canal militated against successful commercial barge traffic and by lending their support to the proposal for the opening up of water communication between Lough Neagh and Lough Erne, and ultimately to the Shannon and Limerick, they hoped that the fortunes of their own concern would be advanced by the creation or encouragement of regular long-distance lighter traffic in heavy, bulky goods and merchandise.[3]

So strong was the current of public feeling that the matter was not allowed to rest and after much preliminary activity a body of landed gentry, merchants, and other public-spirited persons, mainly from the towns and rural areas through which the proposed cut would be made, came together and submitted for parliamentary approval a scheme closely similar to that laid before the Directors General by Killaly some ten years earlier:

> the present is indeed a favourable moment for commencing public works, to give employment both to the peasantry of north and south. Such a measure would do more to tranquillise the disturbed districts of Ireland than can ever be effected by military coercion . . . the utility of a canal from the Blackwater to Lough Erne and its importance to the commerce of Ulster are self evident. That it would afford employment to an immense multitude of the working classes of the community cannot be questioned; and that Irish labourers would be attracted from all parts of the kingdom to the place where they would be sure of obtaining regular and constant work must be manifest to every man who reflects for an instant on the eagerness with which they pass over to England in harvest time to assist in cutting down the corn. It cannot be forgotten that a multitude of Irish labourers were employed in making the Union Canal, which

connects Glasgow with Edinburgh, and are not the coal heavers of London chiefly Irishmen?[4]

However, as the Government was by no means convinced of the certainty of an adequate return from the expenditure of such a large sum of money, the Directors General were powerless to take any practical steps towards beginning canal construction. In view of the pressure of public opinion, however, parliamentary approval was eventually given to the private proposal and a company was incorporated in 1825 to achieve 'the making and maintaining a navigable Canal from Lough Erne in the County of Fermanagh to the River Blackwater, near the village of Charlemont in the County of Armagh'.[5] This was based on a revised survey and estimate by Killaly, which reduced the number of locks to eighteen, and the cost from £223,000 to £160,050.

The Act empowered the Ulster Canal Company to spend £160,000 on the works necessary to establish a line of commercial navigation through Armagh, Monaghan, and Fermanagh; its first list of subscribers included the Marquises of Donegall and Downshire, Lord Rossmore, Sir James Strong and Sir Arthur Chichester. A committee of management was appointed at the first meeting in London (the company being Anglo-Irish), who applied to the Exchequer Bill Loan Commissioners to borrow £100,000, as they were authorized to do by the Act of incorporation. Telford, engineer to the Loan Commissioners, together with their secretary, then went to Ireland and approved the plans and estimates that had been made by the company's engineer, John Killaly. There was then a considerable delay, caused by arguments over the rate of interest to be charged and the amount of the loan, which involved three amending Acts dated 1828, 1829, and 1831, and raised the authority for borrowing from the Exchequer Bill Loan Commissioners to £120,000.

A start was then made with land purchases, and the committee signed an agreement with Henry, Mullins and MacMahon, a firm of Dublin contractors who had subscribed to the company's capital, to build the canal. However, according to an account[6] written in 1832 by an aggrieved Bernard Mullins of the contracting firm, before engineering could be set in hand and 'after a lapse of six years, Mr. Telford discovered that the scale of Lockage he had previously approved of, was unsuitable and that a new survey should be made, for which purpose a deputy of his was sent into Ireland; that no two falls should be combined; that the heights of

ULSTER CANAL,

IRELAND.

A Bill is in progress through Parliament to authorize the formation of a Canal from Blackwater River, at Charlemount, in the County of Armagh, to Lough Erne, in the County of Fermanagh, being a distance of forty-five English miles.

The western part of Lough Erne is about three miles from the Atlantic Ocean, but no safe Harbour exists on the adjacent Coast, or could be made capable of being used by Shipping, excepting during a short period of the summer. By communications from intelligent persons engaged in trade in the neighbourhood of Lough Erne, and in the North Western Ports of Ireland, it has been ascertained that at the beginning of the month of March, there were 40,000 Tons of Grain locked up at those Ports for want of Shipping, which cannot be obtained during the winter months even at excessive freights, and during the summer, the freights, in consequence of the very limited import trade of those Ports, are at least double the rates from the Eastern Ports of Ireland. This is an occurrence of every year. Insurance to a Port in Great Britain is also double the rate of that from Belfast to Newry. The advantage of regularity and a quick market, which may be obtained by means of the Canal, instead of the uncertainty of shipment at the North Western Ports, will command a preference even in summer. The Line of the Canal will pass through or near to the following Market Towns:

DUNGANNON, CHARLEMOUNT, Moy,	ARMAGH, BLACKWATER TOWN, CALEDON,	TYNAN, GLASSLOUGH, MIDDLETON,	MONAGHAN, and CLONES,

and by means of Lough Erne will open a ready communication with numerous other Market Towns in the Counties of Fermanagh, Cavan, Donegal, Leitrim, Sligo, Roscommon, and Longford.

The Blackwater falls into Lough Neagh, from which there are two Canals to the Sea; one to Belfast and the other to Newry.

The survey of a Line for this Canal was originally made under the orders of the Directors General of Inland Navigation of Ireland; it has been since carefully surveyed a second and third time by two different persons, and the present Line, differing only in some very small degree from the original Line pointed out, has been finally adopted by Mr. KILLALLY, one of the principal Civil Engineers employed by His Majesty's Government in the public works now in progress in Ireland.

Mr. KILLALLY's Estimates have been most carefully formed and examined, and there is little doubt that persons of responsibility will contract to do the several works for the sums at which they are estimated.

The Amounts of the Estimates for the whole Line are as follows,—

Earth Work, Excavating and Embanking,...............	£64,290	0	0
Masonry, Seventeen Double and Single Locks............	67,210	0	0
Purchase of Land...................................	14,000	0	0
	£145,500	0	0
Contingencies 10 per Cent...........................	14,550	0	0
Irish Money.......................................	£160,050	0	0
Or Reduced into *British Money*......................	£147,738	0	0

The length of Lough Erne is about thirty-five English miles, the whole of which at an expence of £3000 to £4000 will be open to Navigation, thus throwing into use a Line of Navigable Communication exceeding 80 miles in length, through a most fertile and populous district, at the very moderate expence of about £1850 per mile; a sum considerably under the lowest estimate of any of the recently projected Railways.

The Returns to be obtained from the Canal may be estimated by the fertility of the country through which it is to pass, and which will be opened to a cheap mode of conveyance; by the general want of Fuel throughout the Line of the Canal; by the abundance of Limestone which may be obtained at several places upon the immediate Line of the Canal; and by the facility with which the produce of a large and fertile District, at short distances from the banks of Lough Erne and the adjoining Rivers in the Western part of the North of Ireland, may be brought to Markets at all times accessible to Shipping. The Collieries recently opened near Dungannon promise advantages to the Canal.

Every part of the Canal will be available; as the nearest part to the Town of Belfast is at a distance exceeding forty miles, and is twenty-five miles from Newry. Land Carriage, therefore, cannot compete with the Canal in economy or even in expedition; a Steam vessel capable of towing Lighters being already established on Lough Neagh.

The Supply of Water is ample at all seasons for the most extensive trade.

The Canal may be completed in three years, and an offer has been actually made to complete the works within this period, by a most respectable party, and for a Sum within the Amount of the Estimate.

The following Estimate of the Receipts of Tolls upon the Canal is considered as taken at a very moderate calculation, even in reference to the present state of the country; and does not include Tolls upon any Traffic not commencing at one of the extremities of the Canal.

		Tons.	Miles on Canal.	Rate of Toll.			£	s.	d.
				s.	d.				
From Lough Neagh to	Armagh	2400	4	0	8		80	0	0
Ditto	Caledon	800	9	1	6		60	0	0
Ditto	Middleton	800	12	2	0		80	0	0
Ditto	Glasslough	800	14	2	4		93	6	8
Ditto	Monaghan	2000	18	3	0		300	0	0
Ditto	Smithsboro'	800	22	3	8		146	13	4
Ditto	Clones	1200	30	5	0		300	0	0
Ditto	Lough Erne	16000	36	6	0		4800	0	0
		24800					5860	0	0
Same quantity to Return							5860	0	0
					Total Receipts	£11720	0	0
Deduct Charge of Management, Lock-keepers, Occasional Engineer, &c.						1500	0	0
					Nett Annual Receipts	£10220	0	0

To demonstrate that the above Estimate is by no means over-rated it may be mentioned, that

The Grand Canal, 100 miles long, produces an annual net profit, of Tolls, of £300 per mile.

The Royal Canal, 72 miles long,   ditto   ditto   £120 ditto.

These Canals for forty miles are parallel to each other, and at a distance of only about eight miles.

The rate Toll of the Grand Canal was reduced by purchase of Government, and if there had been only the one Canal for forty miles, there is no doubt that the net profit of Tolls on the Grand Canal would have amounted at least to £350 per mile.

The proposed Cut is thirty-five miles; but the whole new navigation will be about seventy, and to the Sea, including Lough Neagh and the Newry or Lagan Canal, it will be about one hundred miles. Take the net profit of Tolls at two thirds of the Grand Canal, and the net annual income will be £200 per mile or about £14,000 per annum.

The Bill authorises the formation of a Company with a Capital in Shares of £50 each, but the calls upon the Subscribers during the making of the Canal will not exceed £12 10 per Share; His Majesty's Government being authorized under the Act of 3rd Geo. IV. Cap. CXII. to make the further advances in the way of Loan, subject to a low rate of interest, upon the security of the Works to be executed and the Tolls of the Canal, and no proceeding is to be taken towards the formation of the Canal unless the whole number of Shares are subscribed for.

A General Meeting of the Subscribers will be called after the passing of the Act, when they will elect a Committee of Management or Directors, and the Names of Gentlemen of respectability, residents in Ireland, will be presented to the Subscribers for their adoption, if approved, to act in conjunction with the Committee of Subscribers resident in London.

Applications for Shares may be addressed to Messrs. BELL BROTHERS & Co. 164, *Aldersgate Street*, directed "ULSTER CANAL."

A Deposit of £2 10s per Share of £50 to be paid down, and the further calls to be made by the Committee not to exceed £5 per Share, at intervals of not less than three months, as the progress of the Work may require.

The Dividends to be payable and the Shares to be transferable in London.

the several falls should be reduced so that the number of the Locks should be increased from 17 to 26, having falls of from 6 feet to 8 feet 2 inches, and other works of Earth and of Masonry so increased, that the cost of the whole was augmented in the sum of £35,000.' The contractors were then told to submit a new tender. When they did so, Telford told them that his estimate was £15,000 less, and eventually they withdrew from the contract.

John Killaly resurveyed the line in 1831 (he died in 1832), but it is not known whether it was he or Telford who, in order to save money, decided to alter not only the number, but the proposed width of the locks. As we have seen, Killaly's first estimate as to the width of the locks on the Ulster Canal was based on the width of the locks on the Royal Canal, though why he should have chosen this as his criterion remains a mystery. At any rate, even this figure was reduced and the majority of the locks on the Ulster Canal were built some 12 ft in width, up to three feet less than those on the other waterways adjoining Lough Neagh, and so largely preventing through traffic, except by specially built craft.*

Telford himself died in 1834, and was succeeded by William Cubitt. Under Cubitt work proceeded steadily during the remainder of the 1830s, the principal contractor being William Dargan, later to play such a prominent role in the spread of the Irish railway network and in creating the port and harbour of Belfast.

The section at Benburb where the canal, with seven locks in the space of about three-quarters of a mile, was cut into the side of a deep limestone gorge alongside the swiftly flowing Blackwater, gave particular difficulty and several deviations from Killaly's original line were found necessary as engineering proceeded.

the length of this part is about three-fourths of a mile, and it comprises seven locks. The expense of construction, exclusive of the value of land, was £17,053 4 9; in order to diminish the expense as much as possible the canal was contracted in width in two points, where the local impediments were considerable . . . the course of this portion of the line lay along the bottom of a steep ravine in a limestone rock, parallel with the channel of a millrace adjacent to the River Blackwater; the millrace was, therefore, diverted into the river between the first and fifth locks of the canal. Between the third and

---

* Lock widths were as follows:

Newry Canal	15 ft
Lagan Canal	14 ft 10¾ in. to 17 ft 6 in.
Tyrone Navigation	14 ft 10 in. to 16 ft 4½ in.
Ulster Canal	11 ft 8½ in. to 12 ft 6 in.

fifth locks the bed of the canal was formed by benching in the rock on one side and embanking on the other with the materials so obtained; beyond this it was cut for a distance of nearly 350 yards through the limestone, in one place to a depth of 41 feet. The sides and bed were then lined with puddle and protected by a facing of rubble wall. Thence, to the seventh lock, the channel was again formed by benching and embanking through a clay soil where much caution was necessarily exercised in preventing slips at the foot of the embankment which was subject to inundations from the Blackwater. The masonry was all constructed of limestone from an adjacent quarry, the locks being of fine quality ashlar.[7]

By 1838 the cut had been carried as far south-west as Monaghan, the ascent from the Blackwater being achieved by nineteen locks, each with a rise of about 8 ft 6 in. There were landing quays or basins and stores or warehouses at Blackwatertown, Benburb, Milltown, Caledon, Middletown and Monaghan, while the locks and peculiar 'ball alley' bridges were of country rock, mainly carboniferous limestone, quarried in and around Benburb.[8] Between 1838 and 1840 the short summit level between Monaghan and a point short of Smithsboro' was completed. Here was the principal reservoir of the canal, the Quigg-lough (Quigalough) reservoir, a natural lake near the town of Monaghan which, having been embanked, was made capable of holding up to $14\frac{1}{2}$ ft of water, to be released into the canal as required by a regulating sluice and feeder. From the head level the line dropped away towards Lough Erne, passing through Clones and at length entering the river Finn, which ran into Upper Lough Erne, at Wattle Bridge, reached in the autumn of 1841. The descent from the summit level to Upper Lough Erne necessitated seven locks of which the last, at Wattle Bridge, with a width of only 11 ft $8\frac{1}{2}$ in., was the narrowest in Ireland.[9] Construction had cost something over £230,000, of which £120,000 had been granted between 1833 and 1837 as a loan from the Commissioners of Public Works at $3\frac{1}{2}$ per cent interest. By perseverance and engineering skill the lengthy canal had been completed and was now ready to receive its first lighters, be they for Benburb, Caledon, Monaghan, Clones or Enniskillen.

It will be remembered that between 1783 and 1794 work was done by Richard Evans upon a connection between Lower Lough Erne by way of Belleek to Ballyshannon. Killaly subsequently surveyed the area in 1812, and again in 1831, probably in connection with the Ulster Canal. In the following year the possibility of building such a navigation—the main obstacle being the rapids on the Erne below Belleek—was again considered, and Robert Stephenson surveyed the area. He took the view, however, that the cost would be prohibitive, and suggested that a road should be built instead, so laid out that it could be converted to a railway line. Nothing further was heard of the scheme. No trace now exists of the work done by Richard Evans.

As yet there was no canal from Upper Lough Erne to the Shannon, but with the anticipated success of the Ulster Canal, there seemed to be no reason why this cut across the low-lying, lake-studded surface of County Leitrim should not be made in the very near future. Not until this had been accomplished would the Ulster Canal, however successful its own traffic might be, come into its own as a vital link in a great transverse waterway linking Belfast, Newry and the Lough Neagh basin with the Shannon ports and the western seaboard.[10]

For a variety of reasons, however, the Ulster Canal proved an acute disappointment, both to the company who built it and to the many farmers, merchants, and landowners of large areas of Armagh, Monaghan, and Fermanagh who had awaited its opening with eagerness and anticipation. From 1842 to 1851 the canal remained in private hands. With a critical shortage of water on the summit level and in the westernmost section, beyond Clones, enforced transhipment at Charlemont or Blackwatertown made necessary by the narrow locks and shallow channel, and no connecting waterway to the Shannon, it is scarcely surprising that there was no great transfer of traffic from the roads of the area, though the waterway could have provided transport facilities well suited to the bulky traffic in grain, timber, coal, fireclay, pottery, hardware, and provisions which undoubtedly existed at this time in the extensive area served by the new canal. The Ulster Canal Company was quite unable to repay any of the money borrowed from the Government and despite proposals to construct a rail extension from the new Ulster Railway terminus at Armagh, reached in March 1848, to the Ulster Canal at Caledon,[11] barge traffic remained almost negligible.

Eventually, in 1851, the Board of Public Works, as principal mortgagees, assumed control of the canal from the moribund company and after some slight repairs leased it to William Dargan, the contractor, at an annual rent of £400. Dargan had not only built the canal and gained control of its traffic but was also the main carrier, for since its opening in 1842 Dargan's Ulster Canal

Carrying Company had been responsible for moving most of the cargoes moving up and down the waterway. Dargan and the newly formed Lagan Navigation Company were not on the best of terms, so most cargoes from Enniskillen, Clones, and Monaghan during the 1840s were bound for Newry rather than Belfast. Goods and merchandise intended for Belfast were usually transferred to the Ulster Railway at Portadown rather than allowed to move down the Lagan Navigation through Lisburn. Boats left Enniskillen on Wednesday evenings and arrived in Newry on Friday: here cargo was transhipped to cross-channel vessels and could be on sale at Liverpool markets on the following Monday. Freight rates were reasonable by any standards—a firkin of butter from Enniskillen to Liverpool for 1s, a load of timber from Clones to Newry for 3s 6d. Tonnages handled remained very modest, however, and as is usual in such circumstances disuse bred decay in the overall condition of the waterway. The extension of the Ulster Railway from Armagh to Monaghan, reached in 1858, and Clones, in 1863, provided a further dampener to canal traffic and though in 1858 the lease of the waterway was transferred from Dargan to the Dundalk Steam Navigation Company, it made little difference to the fortunes of the navigation, which was by now ruinous. Some idea of the condition of the waterway in 1861 may be gained from the remarks of Sir John Macneill, engineer to many of the principal Irish railway companies. While advocating the further development of railways in the area he remarked that 'the only plan . . . by which any return at all can be obtained from the undertaking . . . is to take off the lock gates, drain the canal and convert its bed and slopes into grassland which may be let for grazing. The banks and wasteland, which in many cases are of considerable width, may be let for tillage'.[12]

Within the relatively short space of twenty years the new waterway had become derelict, a reflection of the folly of building narrow locks, the inadequacy of the water supply, accentuated by imperfect puddle and extensive limestone seepage, the growth of rail competition, the lowering of water levels in Lough Neagh and the Blackwater resulting from the various works of drainage and navigation carried out on the Upper and Lower Bann, Blackwater, and Lough Neagh by the Board of Public Works between 1847 and 1859, and, lastly, of the continued absence of the connecting waterway to the south-west, without which the Ulster Canal lost much of its *raison d'être*.

So serious was the condition of the canal at this time that when the Dundalk company's lease of the waterway expired in 1865, the Government decided, though not without considerable dissent, to assume control of the navigation once more in a final effort to render it fit for commercial traffic. In all probability the completion of the Ballinamore & Ballyconnell Canal in 1860 and the establishing of token lighter traffic between the Shannon and Upper Lough Erne prompted the Government in their decision. At all events, in 1865 the Board of Public Works took over the waterway for a second time, closed it (officially) and in the ensuing eight years spent some £22,000 on works of improvement and overhaul. The greater portion of this was on attempts to secure an adequate water supply for the head level and western reaches from the Quigalough reservoir, but despite extensive works[13] here, in 1874 the reservoir was reported to be completely dry in the summer months.[14]

Reopened in 1873 'tight and in a good condition . . . though liable at any time to give a great deal of trouble' the Board hoped that with drastic improvements to the canal bed and locks and the acquisition of a much more reliable water supply the navigation would at last become a valuable and remunerative artery of commercial transport for large areas of Armagh, Monaghan, and Fermanagh. With the collapse of traffic on the Ballinamore & Ballyconnell Canal after only nine years, the prospect of communication with the Shannon had temporarily receded, but as it was not envisaged at this time that the closure of the Shannon–Erne link was anything but temporary, in 1873 the Board could look to the future with some confidence.

Once again, however, the canal proved a bitter disappointment to the Government, for instead of yielding tolls sufficient to cover the cost of regular works of maintenance and leaving a substantial annual sum for the Exchequer, traffic fell so far short that the Board of Public Works was regularly compelled to expend large sums of money to keep the waterway open. Between 1874 and 1877, for example, annual toll receipts averaged £163 and the annual cost of maintenance approximately £1,250.[15] Public meetings were held at Portadown and Monaghan expressing extreme concern at the condition of the waterway and resolutions were passed calling on the Government to restore the navigation to a condition in which long-distance barge traffic might be reintroduced and drawing attention to the fact that, were this carried out, it would not be difficult to induce local interests to form a company to whose control the canal could well be transferred.

The Government was by now in a state of complete indecision. Having spent a large sum on the canal in an attempt to keep alive the possibility of uninterrupted water communication between east and west coasts and seen their efforts fail, should the Commissioners of Public Works make further attempts to salvage the canal and, by reopening the already derelict Ballinamore & Ballyconnell Canal, attempt to re-establish substantial, long-distance barge traffic in the face of direct rail competition? Should they encourage the formation of local companies to assume control of the several navigations in the area, having previously placed them in a reasonable state of repair? Or should they turn their back on canal communication altogether in view of the continued lack of success, the absence of any definite, foreseeable prospect of improvement and the tightening vice of rail competition?

In an attempt to provide answers to these problems Parliament established two commissions within the space of four years to inquire into navigation works in the north of Ireland, with particular reference to the possibility of reopening the Limerick–Belfast connection.[16] In 1878 forty boats went partly up the Ulster Canal from the Blackwater, paying tolls of some £68, but there was no through traffic to Upper Lough Erne because of water shortage in the summit level and western reaches of the waterway. Indeed, the canal was navigable for only eight months of the year and traffic on it during the remaining four was negligible. From the minutes of evidence of the earlier report it would appear that the Board of Works had already surrendered all hope of servicing the waterway or maintaining it as a navigational link and were intent on disposing of responsibility for it, either by closing it altogether or by transferring control to some unsuspecting private company.

By 1880 the number of boats entering the canal from the Blackwater had risen to 81 and tolls to £87, a slight improvement of perhaps greater significance than might at first seem likely. The probable cause of the increase was that in the interim, and following directly upon views publicly expressed more than once by representatives of the Board of Public Works with regard to the possibility of Government aid being given to a private company showing serious interest in the navigation, the secretary to the Lagan Navigation Company, W. R. Rea, had been sufficiently enterprising to introduce new lighter services on the Ulster Canal, operated by a new carrying company, the Inland Carrying Company, using a number of specially constructed narrow barges carrying 45 tons compared with the 60 or 70 tons carried by craft on the Lagan or the Newry Navigations. Thus, through facilities were provided on the Ulster Canal for the first time for many years. Whereas in 1878 no traffic had reached Upper Lough Erne from Lough Neagh, by 1881 Rea was able to cite the following figures as evidence of the existence of a demand for inland water transport in the area:

To Benburb	630 tons of Indian corn
To Caledon	57 tons
To Middletown	276 tons of miscellaneous merchandise
To Monaghan	902 tons
To Clones	86 tons
To Upper Lough Erne	147 tons (most of this for Enniskillen)

Traffic moving along the Ulster Canal at this time included fireclay from Coalisland (about 100 tons a month, mainly to Monaghan), coal from Belfast and Newry to Benburb, Caledon, Middletown, Monaghan, and Clones, Indian meal from Belfast to Benburb and Belturbet, timber and heavy goods, flaxseed and hardware from Newry to Monaghan and Belturbet. Rea's loaded lighters, moving across Lough Neagh and up the Blackwater, had to be lightened at Charlemont before entering the Ulster Canal, while cargoes arriving in Lagan or Newry Navigation barges had to be transhipped into the special boats of the Inland Carrying Company, either at Portadown or at the entrance to the canal. This caused serious delay to barge traffic and an increase in freight rates, e.g. a cargo of timber moving from Newry to Monaghan took about three weeks in fine weather, up to six weeks in time of frost, at a total freight and tollage rate of about 6s per ton, 2s less than by rail: or again, a cargo of grain from Belfast to Belturbet took well over a fortnight and cost about 9s per ton, as against 12s 6d by rail.[17]

The Royal Commission of 1882 was quick to seize on the interest shown by the secretary of the Lagan Navigation Company in establishing an independent carrying company. Robert Adams, the Board of Works' engineer in charge of the Ulster Canal, stated that in his opinion, after the expenditure of a further £10,000, there would be no scarcity of water on any portion of the canal, while an increase of the depth to 5 ft would quickly give rise to a substantial traffic, which he estimated as follows:

XIX. Lower Bann: (*above*) 'The Cutts' in 1848, site of the lowest lock on the river; (*below*) the staircase (double) lock at Portna, with a flooded dry-dock in the foreground

XX. Lower Bann: (*above*) the sand quays at Toomebridge; (*below*) sand barge from Lough Neagh (formerly owned by Guinness of Dublin), approaching the first lock at Toomebridge

		£	s.	d.
To Benburb	50 tons per week @ 5d per ton	1	0	10
To Caledon	50 tons per week @ 9d per ton	1	17	6
To Middletown	50 tons per week @ 10d per ton	2	1	8
To Monaghan	200 tons per week @ 1s 3d per ton	12	10	0
To Clones	100 tons per week @ 1s 3d per ton	6	5	0
To Smithsboro'	50 tons per week @ 1s 3d per ton	3	2	6
To Upper Lough Erne	400 tons per week @ 1s 3d per ton	25	0	0
	Weekly receipts	51	17	6

In view of this prospect would not the modestly successful Lagan Navigation Company, having already shown a desire to develop feeder traffic entering Lough Neagh from the west, agree to take over the canal provided the Government assisted in its overhaul and improvement to standards acceptable to the Belfast company?

After several unsuccessful attempts to sell the waterway, in lots, to the highest bidder, proposals were at last made by the Board of Public Works to the Lagan Navigation Company suggesting a transfer of the canal. Further communications followed, resulting in the Lagan Navigation Company agreeing to accept it, though with power to abandon it and sell the property if its operation proved unremunerative or impracticable and provided no other solvent company would come forward and accept the responsibility. A bill was introduced by the Government in June, 1884, but was defeated and withdrawn: a second bill was introduced in the following year but was also rejected, while in 1886–7 further attempts by the Government to gain the necessary parliamentary approval for the desired transfer met with as little success, probably due to local mistrust of the Government's intentions. Nothing daunted, the Board of Public Works suggested to the Lagan Navigation Company that it should promote a private motion to the same ends, an astute move rewarded by the successful passage through Parliament in 1888 of the Ulster Canal and Tyrone Navigation Act. By this legislation the Lagan Navigation Company assumed responsibility for the maintenance and operation of both the Ulster Canal and the Coalisland Canal. Unfortunately, however, a committee of the House of Lords declined to accept a clause empowering the private company to close the canal at the end of ten years and, having gone so far, the Lagan Navigation Company reluctantly agreed to accept responsibility for the Ulster Canal in perpetuity. Over the next few years some £12,700 was spent on

H

the waterway, of which the Lagan company provided over £9,000, but though reported in 1896 'to be in fairly good order, in much better order than many similar undertakings', it was soon found that notwithstanding the statement of the Board of Works' engineer in 1882 that £10,000 would be adequate to render the canal watertight and to obtain an adequate and reliable water supply, during the summer months in particular water was quite insufficient and regular traffic impossible.

The Lagan company was thus saddled with an assured annual liability, for although annual toll receipts increased from approximately £150 in the 1870s to over £700 in the 1890s, upkeep and maintenance of the navigation still cost between £1,200 and £1,400 every year. Indeed, more often than not the revenue from the Lagan and Coalisland Canals was swallowed up by this regular deficit on the Ulster, so much so that the company's annual dividend declined rapidly from an average of $1\frac{1}{2}$ per cent over the period 1877–82, to the following:[18]

Year	Rate of dividend declared per cent	Year	Rate of dividend declared per cent
1890	Nil	1896	Nil
1891	$\frac{1}{2}$	1897	$1\frac{1}{4}$
1892	Nil	1898	Nil
1893	Nil	1899	1
1894	Nil	1900	1
1895	$\frac{3}{4}$	1901	$\frac{3}{4}$

With the acquisition of the Ulster Canal, a waterway which no reasonable amount of money could have rendered navigable and remunerative, the Lagan Navigation Company, which had hitherto enjoyed a fair measure of success, was threatened with extinction. It was a blow from which it never really recovered.

An analysis of barge traffic on the Ulster Canal over the period 1890–1905 shows that while traffic and toll receipts increased fairly steadily, they still remained well short of the amount expended on normal works of maintenance and day-to-day operation. The result of lack of standardization in the locks of the various navigations was that boats closely adapted to the gauge of the Lagan Canal could use the Upper Bann and Blackwater and, with lightened cargoes, the Coalisland Canal, but not the Ulster Canal. Had they been able to negotiate that waterway they could have used Upper Lough Erne and served Enniskillen but at this time such journeys were quite impossible. Without enlargement

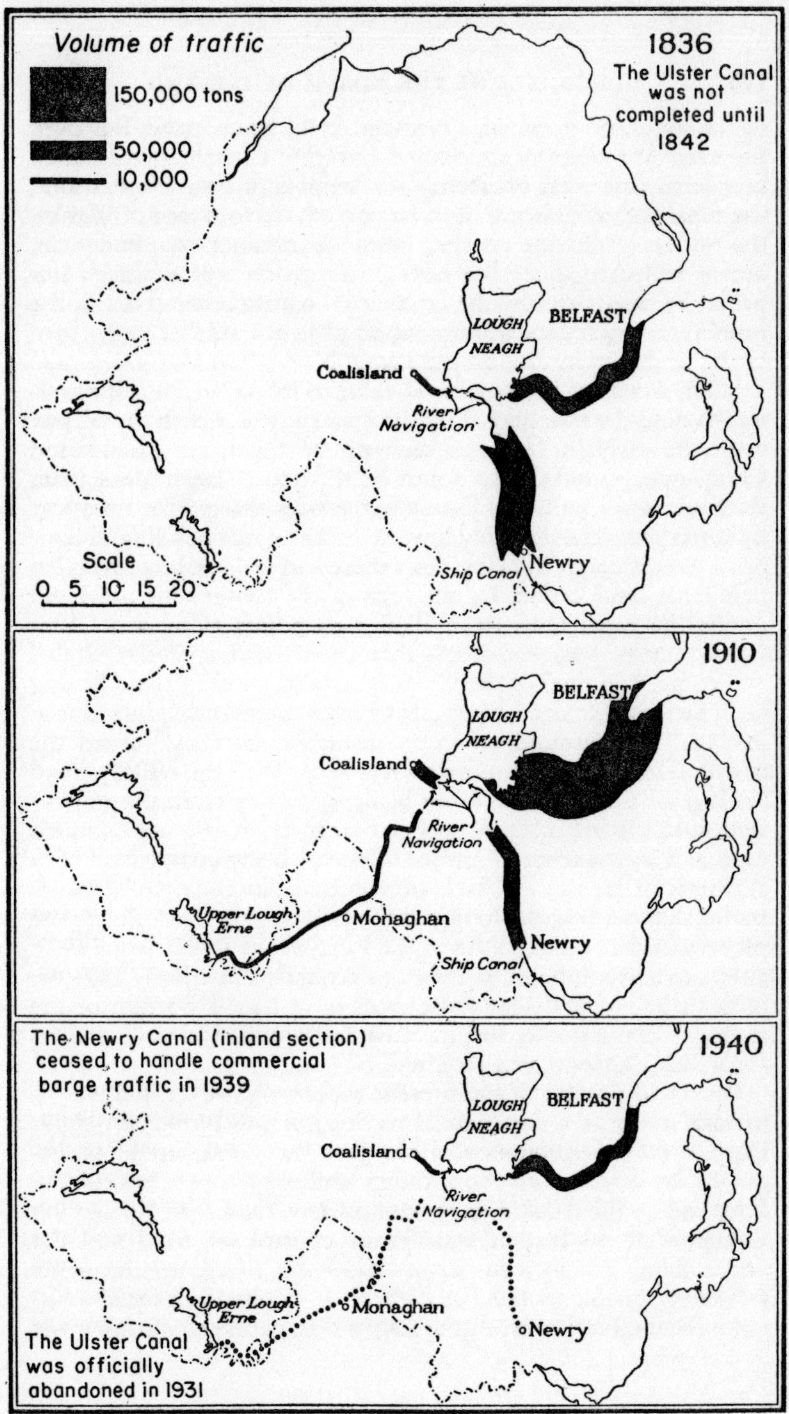

14. Flow-line diagram illustrating traffic on four of Ulster's principal canals

of the locks, no great improvement could be expected, but even the narrower boats of the Inland Carrying Company were often unable to pass west of Monaghan because of insufficient water, the canal bed often being quite dry for two to four months during the summer. On the section from Charlemont to Monaghan, which still carried regular traffic, navigation was so much impeded by weeds and by the constricted nature of stretches of the canal that two horses were required to haul a smaller cargo than could be drawn by one on the Lagan Navigation.[19]

While it was of little practical value to its owners, it is interesting to note the role played by the waterway as a curb on railway rates. As early as 1882 the secretary of the Lagan Navigation Company reported that to points on the Ulster Canal, Monaghan, Benburb, and Clones, for example, the rates charged for transport by water were considerably lower than the corresponding railway rates, even though in many cases these had been reduced to compete with canal traffic. By the turn of the century for nearly all goods the canal rates from Belfast to points along the Ulster Canal remained somewhat less than those existing on the parallel and competing lines of the Great Northern, most of these having been increased under a new railway rates classification introduced in 1893. Furthermore, in many instances the canal served the towns along its course more conveniently than the railway, road haulage of up to a mile often being necessary from the railway station to the town centre and market place. It was not so much rates and convenience of service that were being complained of as slowness of transit and lack of regularity in commercial barge traffic. Although in the long run the railway was to accelerate the process of decay and disuse of the canal as yet its existence as an alternative was enough to enforce a reduction in railway rates or, rather later, to undercut these rates in at least a portion of the traffic in heavy goods and merchandise, grain, fuel, and timber, for which transport was required.

By the beginning of the present century, however, the end of the waterway as a commercial navigation was already in sight. Despite official recommendations that the canal should be acquired by a proposed 'controlling authority' on behalf of the State and 'to the relief of its present owners, the Lagan Navigation Company',[20] no further transfer of control occurred and the undertaking remained in private hands for the remainder of its existence. Unable to transfer control or to obtain financial assistance from official sources, the Lagan Navigation Company made several attempts to secure legislation authorizing the abandonment of the canal as a commercial waterway, but though a bill was introduced at Westminster on 23 October 1915 to permit the temporary closing of the canal, it was so vigorously opposed by local interests that it was later withdrawn. From this time to 1929, the amounts of merchandise and fuel being carried along the waterway steadily lessened and the annual trading deficit widened to such an extent that for six years between 1916 and 1931 the Lagan Navigation Company showed a clear deficit on the working of the three canals under its control in the north of Ireland.

The last lighter entered the canal on 29 October 1929, and on 9 January 1931, the company succeeded in obtaining an official 'warrant of abandonment' for that portion of the canal that lay in the 'Six Counties' from the Ministry of Commerce in the Government of Northern Ireland. On 15 April an 'order of release' relieved the company of all liability in respect of maintenance and vested the property in the parties who had previously purchased the various works (bridges, locks and lock-keepers houses, sheds, quays, etc.) and sections of the canal bed and towing path from the navigation company. In Eire the canal reverted to the Ministry of Industry and Commerce, successors in the Irish Republic to the Board of Public Works, who subsequently sold the lock-houses and former canal bed, mainly to the tenant lock-keepers.

The Ulster Canal was an unfortunate undertaking from its conception in the early years of the nineteenth century to its final closure little over a century later. Its construction was prolonged and fraught with difficulties, its early administration left much to be desired, its water level was always defective, its draught too shallow, its locks too narrow. It was in competition with rail transport for the greater part of its existence and soon became an item to be disposed of by its controlling body, firstly the Board of Public Works and, for the last half-century of its existence, the Lagan Navigation Company.

It was conceived as an important and extensive link in the line of inland water communication to the Shannon connecting Limerick, Athlone, and Carrick-on-Shannon with Belfast, Newry, and Coleraine through Lough Neagh, which would both supplement and compete with the Grand and Royal Canals farther south. In this it suffered by the failure of the Ballinamore & Ballyconnell Canal, linking Upper Lough Erne with the Shannon. It was in all probability only because of the remote possibility of this west-east water connection proving practicable and being reopened that the

Ulster Canal was maintained at all after 1880. Of itself it had little commercial significance after that date, and was a steady drain on the finances of the Lagan Navigation Company. Apart from acting as a temporary curb on railway rates, indeed, it was never of great practical utility to the areas through which it passed and after its division in 1921, when it lay partly in Northern Ireland and partly in the Irish Republic, neither of whose Governments was particularly anxious that it should be maintained, its formal abandonment was merely a series of intricate legal steps with but characteristic local opposition to be circumvented.

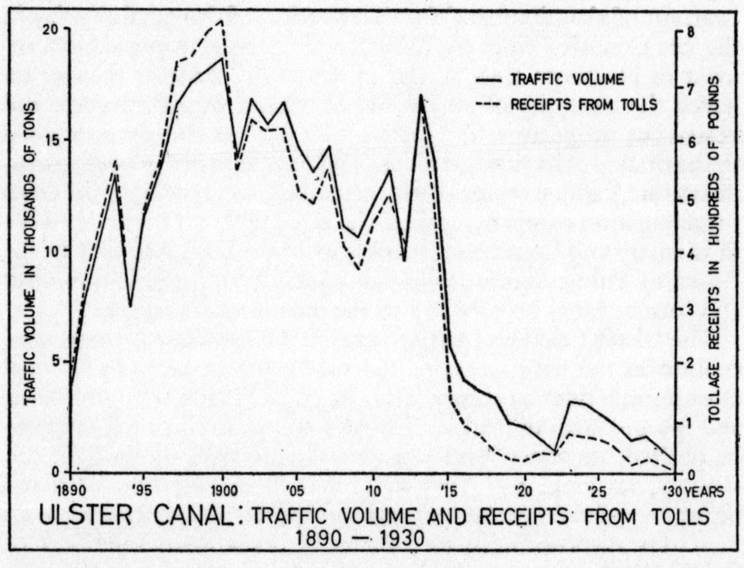

15. Ulster Canal graph

It was unfortunate for the Lagan Navigation Company that it was induced to accept control of the waterway towards the end of the nineteenth century and in view of the circumstances attending that transfer, it does seem, in retrospect, as though subsequent treatment meted out to the company by the Board of Public Works was unduly harsh. It was not the fault of the Belfast company that the waterway remained a dismal failure: as it existed, indeed, the canal had little chance of success, regardless of who controlled it. Serious natural difficulties and physical disadvantages were the real causes of its lack of success, a backcloth against which factors of economic expansion, commercial development and company administration interacted one with the other to produce varying degrees of disappointment and failure.

CHAPTER VII

The Lower and Upper Bann and Lough Neagh

THE only outlet of Lough Neagh, the largest freshwater lake in the British Isles, is the Lower Bann which enters the sea below Coleraine. The Blackwater, the Ballinderry, the Moyola, the Upper Bann, the Sixmilewater, and the Main, with many small tributaries, flow into it, while the catchment area drained is over 1,500 square miles. The lough has always acted as a natural reservoir, but in spite of its extent of 153 square miles, it is relatively shallow and therefore its capacity is not as great as might be expected. In past years the inability of the lough to absorb the surplus of inflow over discharge has caused serious flooding over a large area round its shores, and it has been estimated that in its natural state, prior to the mid-nineteenth century, some 25,000 acres were subject to regular inundation.

As long ago as 1738 the Irish Parliament discussed a project for improving matters by removing a rock barrier on the Lower Bann at Kilrea (Portna). The Bishop of Down and Connor, Francis Hutchinson, who then lived at Portglenone, was the moving spirit in this attempt to alleviate the widespread flooding. According to Hutchinson[1]

> the waters which flow from so many sources can not possibly be discharged by the single outlet of the Bann but must, unless steps are taken to discharge of the waters by clearing the obstructions of the river, be annually accumulated to the great detriment of the lands around.

Commenting on the effects of the regular overspilling of the waters of Lough Neagh the Bishop remarks that

> Ballyscullen church was not only encompassed by the floods but a great part of the parish had been drowned, those great tracts of rich

land, once adorned with trees, were covered and a fisherman, having twice removed his habitation, was about to do so again, complaining that he knew not where to place it, for the Bann followed him.

Eventually the Government made preparations to lower the level of the lough by drainage, but the Bishop died and nothing was then done.

Lower Bann Navigation

Flooding remained a constant problem during the next hundred years and many proposals were made for drainage, with or without provision also for navigation.[2] Of particular interest was the scheme outlined by Alexander Nimmo in 1822. On being consulted by parties interested in improving the Newry Navigation, he stated that

> this navigation i.e., the Newry, is capable of a singular improvement, the effect of which would be of the greatest public consequence, not only to the trade of Newry, but to all the country around Lough Neagh . . . it would be a matter of no great expense or difficulty to cut the summit level of the Newry Inland Canal to a depth sufficient to bring the surface level of Lough Neagh through the valley towards Newry, giving a supply of water to the canal and our proposed lower navigation that would be abundant and inexhaustible, saving also the passage of three locks on one side and four on the other. I should propose to bring that level on to the town, which the ground readily admits of, and thus the whole of the inland trade of Newry may be done without lockage at all and the cutt may be of such dimensions as to admit of steam vessels, so as to excel any kind of land carriage, even in expedition.
>
> A slight improvement on the Lower Bann would extend this level navigation to Portglenone, sixty miles, in a straight line, from Newry. We could obtain a command of Lough Neagh such as to run it down some feet in summer and keep its winter rise to the present summer level, thus saving from being flooded six thousand acres of land. We would also bring a water power into Newry of, say, one-sixth of the discharge of the lake, with forty-five feet of fall, which I estimate at three thousand horse power, being more than that of all the steam engines of Glasgow. The reservoir covers one hundred thousand acres and the value of such a power in a seaport and in the midst of a rich, populous and busily manufacturing country, is not to be easily calculated—indeed, the like is not to be found in Europe. . . .[3]

However, despite such ambitious proposals as these, the first definite step towards effecting an improvement in the condition of the lands around Lough Neagh and in promoting inland navigation in the area took place in 1842 when the Imperial Government, headed by Sir Robert Peel, succeeded in passing a bill for the improvement of drainage, navigation and mill power in the area, the execution of the works being made the responsibility of the Government, who were also given power to assess the cost of drainage improvements on the landowners and of navigation works on the ratepayers of the areas concerned.[4] In the north of Ireland the initiative was taken by Lord Lurgan who owned some twenty-one miles of shoreline along Lough Neagh and the Upper Bann and had spent thousands of pounds on embankments, drains, roads, and pumping engines, all without success. In the spring of 1844 he had a memorial signed by the principal landowners in the area asking that the level of the lough should be lowered and that navigation of the waterways of the neighbourhood should be improved, lodged £1,500 for preliminary expenses and presented a petition to the Commissioners of Public Works in Dublin:[5]

> Your memorialists are proprietors of and parties interested in certain tracts of land adjoining Lough Neagh in the counties of Antrim, Down, Armagh, Tyrone and Derry, which lands are generally of an alluvial nature and are supposed to contain about 20,000 acres and extend all round the said lough and its rivers; which lands are flooded or otherwise injured by water for about six months in the year and are capable of being drained and improved. The keeping of Lough Neagh at summer level would prevent those serious evils which occur to the health of the community and those obstacles placed in the way of improved agriculture by the winter floodings of Lough Neagh, conjointly with a vast improvement of the navigation of the district.
>
> Your memorialists are desirous of availing themselves of the provisions of the Drainage Act recently passed for the purpose of relieving the said lands and such others as your Award may consider worthy of relief by the necessary work from flood and other injury by water. Your memorialists therefore pray that all the lands so flooded may be drained and the drainage of all the lands so injured may be improved, in conjunction with the improvement of the navigation of the district, so far as the improvement of the said navigations forms a part of cheap and efficient drainage under the provisions of the aforesaid Act.

In November 1845, another memorial, signed by both the Protestant and Roman Catholic Primates, besides Bishops of both churches, numerous clergy and magistrates, urged that, in view of the failure of the potato crop, employment should be given to the labouring population on works of permanent and practical utility, of which the Lough Neagh scheme and the navigation connected with it were 'by far the most important'.

Already, in the summer of 1843, a detailed survey of the Lower Bann with a view to its conversion to a navigable waterway had been carried out by the British engineer, Francis Giles, assisted by James McCleery, engineer and secretary to the Lagan Navigation Company. On 8 January 1844,[6] a 'numerous and respectable meeting of the gentry, merchants and citizens' of Coleraine was held in the Courthouse at which McCleery read a report of the contemplated undertaking, giving £41,000 as his estimate of the works that would be required on the river to make it navigable. The proposed scale of the navigation and its locks, with the exception of that at the Salmon Leap near Coleraine, was to be the same as on the Ulster Canal. Shortly afterwards, however, a prospectus was circulated giving details of the engineering survey by Giles and McCleery and inviting subscriptions towards the formation of a company 'for carrying the navigation measure' into execution, and in this the estimated cost was given as £25,000.

The first official move towards establishing a navigation along the Lower Bann and in Lough Neagh occurred in 1846 when John McMahon, a government engineer, conducted an exhaustive inquiry into the entire problem and presented a most detailed report.[7] McMahon pointed out that intake from the rivers falling into Lough Neagh, the Lough itself, and the basin of the Lower Bann, some 340 square miles, all had to be discharged by one obstructed outlet. The inevitable flooding rendered some 25,000 acres profitless.

He recognized the great benefits bestowed by the large expanse of inland sea, even in its natural state, in the control and storage of water: 'placed by nature at a point of convergence of several powerful and turbulent rivers and streams, it receives and calms the impetuosity of those waters, rendering them fit for man's use and is almost without a parallel as to value amongst his industrial resources'. Yet the problems were great, for the river Blackwater alone discharged 500,000 cu. ft per minute into the lough during flood, while the maximum discharged down the Lower Bann he put at only 400,000 cu. ft per minute, without taking into account another 1,248 square miles of land, the excess water from which had also to find its way down that river. He proposed to reduce the level of Lough Neagh to the summer level of 1826, the lowest known for many years and about six feet below the usual surface level, and to limit the possible rise to one foot. The lower sills of the navigations connected with Lough Neagh, the Lagan, Newry, Ulster, and Coalisland Canals, were also to be reduced and the bars at the mouths of the Upper and Lower Bann cleared by dredging. The surface of Lough Beg and a stretch of the Lower Bann down to Portna (Kilrea) were to be lowered 'in order to afford sufficient capacity in time of flood for the discharge of the surplus waters'. At Portna, some fifteen miles below Toomebridge, a transverse ridge of solid rock which McMahon recognized as a major obstacle in solving drainage problems in the entire Lower Bann valley formed a local base level, raising the whole bed of the river above that point. This was to be excavated and a weir constructed in addition to a lock and side channel for navigation. The course of the Lower Bann was to be cleared, the waters regulated by weirs and locks, and a navigation established 'ample for the transit of screw steam trade boats of considerable tonnage to the town of Coleraine, and thence by the tidal waters to the sea'. In doing so water power would be developed and it was anticipated that a large revenue would be obtained from the mill sites. Considerable expenditure was also to be incurred on two bridges across the Lower Bann, at Toome and Portglenone.

The most glowing anticipations of the benefits to be derived from this scheme were entertained, doctors reported upon its probable effect in abating the epidemics of fever; professors of geology discoursed on the economic value of the deposits of clay which would be exposed near Toome and farther south between Arboe Point and Portadown—diatomite, peat, lignite and coal, sulphate of lime, and sulphate of iron; the Drainage Commissioner, Robert Harding,[8] contemplated a vast improvement of the land which was liable to flooding. In short, nearly everyone in the Bann corridor, both north and south of Lough Neagh, was to be benefited in one way or another.

On the Upper Bann river, McMahon proposed the excavation of shoals at Bannwaterfoot, Portadown and upstream as far as Knock Bridge, at an estimated cost of £6,000. The river would quickly improve with the lowering of the level of Lough Neagh and the clearing of the course of the Lower Bann. The works which he contemplated on the Blackwater were as follows:

(*i*) the reconstruction of the embankment and sluices above Charlemont Bridge, downwards to Verner's Bridge, up the

Callan and Tall and up the Blackwater to the junction with the Ulster Canal and the increase of the gradient of the main stream from Charlemont to Lough Neagh to nine inches per mile—estimated cost £4,210;

(*ii*) the excavation of shoals, the straightening of the river's course and the general deepening of the channel—estimated cost £9,600;

(*iii*) the strengthening of the foundations of Verner's Bridge—estimated cost £190;

(*iv*) the removal of the old bridge at Charlemont, the new bridge to cost £3,000.

In April 1846 a 'schedule' on the drainage section of the entire Lough Neagh scheme was submitted by Robert Harding,[9] giving the lands which would be drained and improved, setting out the names of the owners, the area concerned, its value at the time and an estimated increase in value due to the proposed improvements. The total cost of the drainage works was estimated at £109,500, to be paid by the land owners of the areas affected.

In May 1846 another 'schedule' was published,[10] again prepared by Robert Harding, setting out the districts to be benefited by the navigation works 'and the proportions in which the same are likely to be benefited and should contribute to the proposed works of navigation'. These were estimated at £74,275.

Thus, the estimate for all the drainage and navigation works was £183,775, of which £109,500 was to be raised by a proportionate levy on all landowners and £74,275 by the areas to be served by the navigations, in proportion as each was likely to be benefited.

McMahon's report having been made and detailed financial schedules for the works prepared by Harding, Lord Lurgan succeeded in obtaining a grant equivalent to half the cost of the navigation works, a figure ultimately increased to two-thirds on the grounds that, being relief works, more was spent than was really necessary. Numerous meetings called by interested parties were held in the months following the publication of McMahon's plans and Harding's schedules. Eventually, the required assent of the majority of landowners was obtained and, all the requirements of the Act of 1842 having been complied with, the works were begun in 1847, under Charles Ottley, C.E.

The engineering involved occupied the next ten years and cost a great deal more than the original estimate, a total of £254,167 up

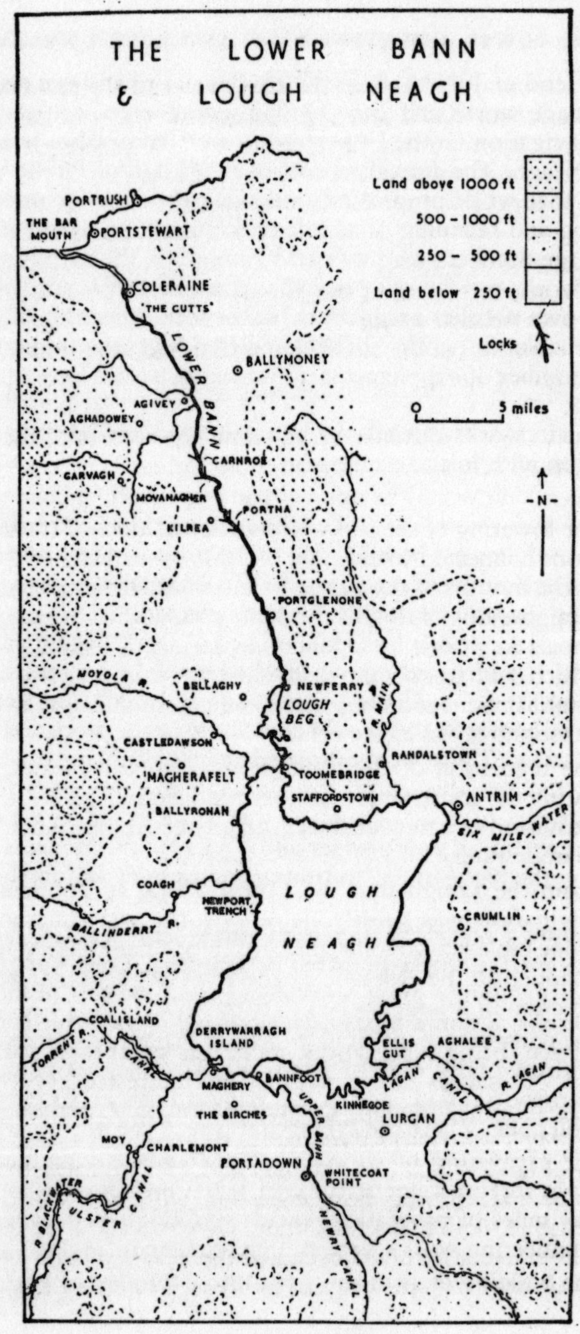

16. Map of the Lower Bann, etc.

until the end of July 1858, of which £144,214 had been spent on the drainage works and £101,081 on works aimed at promoting inland navigation on the Upper and Lower Bann, Blackwater and Lough Neagh. The annual reports of the Board of Public Works indicate that work progressed under increasing difficulties—the procuring and retaining of an adequate supply of labour in view of the large numbers who wanted to emigrate, the carrying on of railway works not far away and the generally ample employment 'in their own peculiar occupations' of the local labouring classes—and to enable the works to be completed with the minimum of delay a number of supplementary Acts were passed between 1847 and 1858.

The main works carried out between 1847 and 1858 in direct connection with inland navigation were:

(*i*) the lowering of the sills of the entrance locks of the navigations connected with Lough Neagh, the cutting of the bars at the mouths of the Upper Bann and Blackwater rivers, the straightening of the Blackwater's course and the removal of shoals in it and, by a general lowering of the water level, which had frequently rendered river banks invisible, the establishment of an uninterrupted communication by water throughout the entire area;

(*ii*) the regulation of the water level in Lough Neagh, facilitating the construction of quays and landing places, and the provision of a new quay at Antrim near the entrance of the Sixmilewater;

(*iii*) from the Lough into the Lower Bann, at Toomebridge, navigation was greatly helped by the construction of a stone mole 800 yd long, the cutting of a flanking navigation channel on the east side of the river at this point and the erection of a large lock, 130 ft by 20 ft;

(*iv*) the construction of similar locks at Portna (a staircase lock with two falls, each of 7 ft 3 in.), Movanagher 10 ft 6 in., Carnroe 6 ft 6 in., and at the Salmon Leap or 'Cutts' above Coleraine, where the average fall was 6 ft 6 in.;

(*v*) the excavation of shoals and sandbanks giving seven to eight feet of water at summer level throughout the entire $32\frac{3}{8}$ miles of navigation, except in Lough Beg, where the engineering works were still incomplete in 1858;

(*vi*) the building of swing sections into a number of bridges.

It was anticipated that

> in view of the cheap and facile means of transport which the extended navigation affords, as well as the local quay accommodation which has been provided for the thriving towns and improving agricultural districts adjoining the Lower Bann, there would develop a considerable trade in heavy goods, such as bricks and tiles, coals, lime and building stone, grain and timber, of which in fact there are at present a very satisfactory indication, and a most important benefit will accrue to the surrounding country but more especially to those portions of it which are included in the declaration schedule as liable for a moiety of the outlay.[11]

As the assent of the landowners had only been obtained for an expenditure of £109,500 on drainage works, the Commissioners of Public Works, under whose auspices the scheme had been carried out, recommended that these proprietors should not be called upon to pay the excess (£34,714) and the Treasury agreed. With regard to the navigation works, while the ratepayers were legally liable to pay half the cost, the Commissioners recommended that their liability should be a moiety of Harding's original estimate of £74,275 rather than half the actual expenditure of £101,081. Here again the Treasury gave its consent and, indeed, the Government provided another £2,000 towards dredging an adequate navigation channel through Lough Beg.

The works having been completed, the findings of the Commissioners of Public Works upon the financial schedules, which kept closely to those originally drawn up by Harding in 1846, were made known late in 1858[12] and a public inquiry was held in Belfast soon afterwards.[13] To offset the expenditure incurred on navigation works between 1847 and 1858, a charge was to be made on each county as follows:

	£	s	d
Antrim	15,032	10	4
Londonderry	12,894	17	8
Tyrone	6,128	1	2
Armagh	3,018	3	3
Down	63	17	7
	37,137	10	0

For this sum, therefore, the areas bordering Lough Neagh or the lowland corridors to north and south received navigation works which had cost over £100,000. Having completed the works of navigation and drainage, the Commissioners of Public Works in

1859 transferred responsibility for their maintenance to three distinct bodies, the Upper Bann Navigation Trust, the Lower Bann Navigation Trust and the Lough Neagh Drainage Trust, whose authority also extended to controlling and effecting drainage in the catchment area of the Blackwater.

Whereas by Harding's original schedule of 1846, responsibility for the entire navigation was to be spread over the five counties benefiting from it, by 19 & 20 Vic. c. 62 (21 July 1856) the works situated on the Lower Bann and in the northern waters of Lough Neagh became the responsibility of Counties Antrim and Londonderry alone while the remainder passed to Counties Armagh, Tyrone and Down. With the appointment of a Lower Bann Navigation Trust, an Upper Bann Navigation Trust, and a Lough Neagh Drainage Trust in 1859, this meant that once the cost of the navigation works had been paid off, any deficiency in annual working of the Lower Bann Navigation was to be made up by the ratepayers of Counties Antrim and Londonderry and similarly for the Upper Bann Navigation by Counties Tyrone, Armagh and Down. The proportions in which they were to make good any deficiency were to be the same as those used for the repayment of the principal sum.

High hopes had been expressed in the early 1840s for the extensive drainage and navigation schemes that had been proposed but we cannot judge whether the expenditure of a very large sum of public money, in addition to a substantial annual levy over large areas of mid-Ulster, would have been justified by results, for while the work was being carried out, the situation had been completely altered by the development of a railway system throughout the Lough Neagh basin.

In 1846 the only town in the district served by rail was Portadown. William Dargan, under whose direction the line from Belfast to Portadown had been constructed, owned lands on the shore of Lough Neagh and had been one of those who had signed the memorial of 1844 asking that the drainage scheme be undertaken. To develop traffic on the newly constructed railway he put a steamer* on Lough Neagh, which operated principally between

* It was said that a look-out man on the roof of the Lurgan church watched the sailing of the *Grand Junction* paddle steamer from Newport Trench or Ballyronen through a telescope and that the Captain would signal the number of passengers on board who would have to get conveyances to go to fairs in County Armagh. The anecdote is more likely to refer to the *Lady of the Lake*, a paddle steamer belonging to James Gaussen of Ballyronan, which ran regularly from Ballyronan to Kinnego in the late 1830s, but was withdrawn when the Ulster Railway reached Portadown and Dargan introduced his service.

I

Newport Trench and Portadown, carrying passengers, goods, and livestock, a service which handled a large traffic during the famine years. Emigrants leaving the poorer districts west of Lough Neagh boarded the steamer at Newport Trench or Ballyronan for Portadown or Lurgan (Kinnego) where they could join the Ulster Railway to Belfast on their way to England or America.

In the spring of 1848 the Ulster Railway to Armagh was completed, while shortly afterwards the Northern Counties line was opened to Randalstown and Ballymena. Before the works on the Lower Bann were finished and the contemplated steamer service introduced, the northern railhead was driven on through Ballymoney to Coleraine, with a branch line to Portrush, reached in November 1856. Twelve months later the more southerly railhead, crossing the river at Toomebridge, reached Cookstown through Magherafelt and Moneymore. Thus potential traffic in the Lower Bann corridor was already effectively served by railways even before the navigation works were ready to handle their first vessels.

To the south of Lough Neagh similar developments were occurring. By April 1858, the railway had been extended from Portadown across the Blackwater to Dungannon, while by August 1861 it had been driven across the watershed of the moorlands of the south Sperrins through Pomeroy and Carrickmore to Omagh at the head of the Foyle drainage system.

Whether this programme of rapid railway extension was directly influenced by potential competition from the proposed navigation along the Lower Bann and on Lough Neagh is difficult to say, but it had the effect of severely blighting the prospects of the Lower Bann as an artery for traffic from the inland sea to and from the north coast. So marked was the effect of this rapid railway development that by 1856 the prospects of commercial success for the navigation had deteriorated to a point where there seems to have been little or no objection to the passage of an Act severing the navigation works and handing those on the Lower Bann to the two counties bordering that river and those around Lough Neagh and on the Blackwater and Upper Bann to certain areas of Armagh, Tyrone, and Down; this in spite of the fact that only ten years earlier the glowing anticipations expressed by McMahon with regard to the success of the navigation works would have made it highly improbable that such a step could have been taken without a major public outcry.

Subsequent rail extensions contributed further to the difficulties of commercial navigations in the Lough Neagh basin. In 1871 the line from Lisburn to Antrim was opened while in 1879 Cookstown was linked to the Great Northern at Dungannon through Stewartstown and Coalisland. Lough Neagh was now entirely surrounded by railways and with the opening in February 1880 of the Derry Central Railway Company's line, running almost parallel with the Lower Bann on the western side, the commercial success of the Lower Bann Navigation was rendered even more improbable.

Although an improvement in the general drainage of the Lower Bann area did occur during the years immediately after the completion of the works in 1859, there was subsequent deterioration and flooding again became a serious menace. However, it would be unfair to represent the works carried out between 1847 and 1859 as a failure. On the contrary, with the exception of large-scale navigation on the Lower Bann and the development of mill power, they were as successful as could reasonably be expected. A large area of land was reclaimed, navigation was for a time developed, and at least some of the secondary anticipations were realized.

One important drainage difficulty was that every acre reclaimed and drained provided additional run-off to be disposed of down the main stream below Toomebridge and this, coupled with the unreliable nature of the data on which the works had been constructed and the apparent ignorance of the Lower Bann Navigation Trust of their responsibilities in maintaining a clear, uninterrupted channel in that river, did much to nullify the good intentions behind the original Act of 1842.

Nor did navigation on Lough Neagh or along the Lower Bann ever approach the dimensions envisaged in the 1840s. In 1846 McMahon had noted that

the Lower Bann, the drainage conduit in which the navigation is intended to be carried for its whole length, thirty-seven miles, affords ample breadth and depth of water for a most useful class of steam propelled steamboats, with powers sufficient to stem the currents of the river, the violence of Lough Neagh gales and the tides of the Bann estuary. Such boats as tugs will also greatly aid the cross-channel trade with Coleraine and with the interior of the country when a channel connecting the navigation with the river below the bridge shall be completed.[14]

Dealing with the general prospects he continued:

> the country through which the navigation is carried on is populous, its trade thriving and its land productive. There are several rising towns on its banks, and contiguous to its course, and the terminus, the town of Coleraine, is prosperous and public spirited. If then, as is herein proposed, the waters of Lough Neagh be united to the port of Coleraine by a commodious navigation, the enterprising Coleraine merchants will have opened to them the market of a large share of the produce of those districts impinging on the lough, which are now, from the geological configuration of the country, quite beyond their reach.[15]

These benefits did not appear, for neither the drainage scheme nor the navigation works on the Lower Bann, as they had been carried out by the Board of Public Works, proved satisfactory. The original drainage scheme was made less effective by the navigation works, and the anticipated flow of commerce between Coleraine, towns on or near the Lower Bann and ports around Lough Neagh, never materialized. The dual project had fallen between two stools, failing to give satisfaction to riparian landowners on whom a heavy drainage tax was levied and at the same time failing to create substantial freight traffic between Coleraine and the area which it was hoped would become an elongated hinterland.

By the 1870s the annual receipts of the Lower Bann Navigation Trust were a mere £100, while the annual cost of maintenance often exceeded five times that figure.[16] Besides slight barge traffic in tug-drawn trains, a steamer, the *Kitty of Coleraine* was put on the Lower Bann for a time to carry passengers and goods, but it was unsuccessful because when fully laden it could not move upstream against the strong current. It was soon withdrawn and replaced by a more powerful vessel, the *Banshee*, but this too, was unsuccessful, for she drew more water than was usually available. With the withdrawal of the second vessel, prospects of establishing regular and substantial commercial navigation on the river seemed bleak indeed.[17]

For a time, however, it seemed as if external developments might temporarily stimulate freight traffic on the Lower Bann, at a time when the navigation was showing a substantial annual loss and was in danger of abandonment. Until now, Portrush had been the main port for cross-channel traffic to and from the Lower Bann valley and Lough Neagh, because, owing to a sand bar at the mouth of the river, only small vessels could moor at Coleraine.

The result was a costly and time-wasting transhipment from steamer to railway truck for haulage to Coleraine, Kilrea, Garvagh, Maghera and Portglenone.

In 1880 the Coleraine Harbour Commissioners authorized the erection of two moles at the mouth of the Lower Bann, harbour facilities near the town centre, and a general dredging of the tidal estuary, at an estimated cost of £88,000, in the hope of capturing most of the traffic passing through Portrush. It was natural that the Coleraine port authorities should favour the development of navigation on the Lower Bann and Lough Neagh, for barge-trains drawn by steam tugs could distribute goods brought to the new cross-channel port, and collect agricultural produce from a large area of mid-Ulster for shipment at its quays.

Coleraine did, in fact, rob Portrush of much of its trade and develop as the major port of the Lower Bann area, with a regular cross-channel service to Glasgow from 1884 onwards. On the other hand, the building of efficient rail links with Belfast from Cookstown, Magherafelt, Toomebridge, and the towns of the Lower Bann valley, as well as from Ballymena, Ballymoney, and Coleraine itself, did much to direct the trade of the Lower Bann more and more through the great east coast port. The ability to dock larger vessels at Belfast, cheaper cross-channel rates and the relative ease and efficiency of rail and road communication between the Lower Bann area and the Lagan valley, both north and south of Lough Neagh, all contributed to an acceleration of the spread of Belfast's trade into the area with which we are concerned.

Further, the commercial waterways of the Lough Neagh basin came late in the pattern of development of inland navigation in the north of Ireland. Unlike those canals which had established a considerable traffic during the first half of the nineteenth century, particularly the Lagan Navigation, the Lower Bann, opened as late as 1859, had failed to attract much commercial traffic before the full force of competition from the railways, and from the increasing predominance of Belfast, came to be felt. By the 1880s its maintenance and operation were resulting in an annual deficit of over £1,000. It was scarcely likely that the position would improve as the century drew to its close and competition from rail transport became more intense.

However, the navigation did remain in operation, despite the recommendations of Royal Commissions of 1882 and 1887. The former, 'enquiring into the system of navigation which connects

Coleraine, Belfast and Limerick',[18] concluded that

> if the counties who contributed to the cost of the works consent, the Board of Navigation Trustees should be dissolved and all the works, with the entire control of the river, should be transferred to drainage trustees, with ample powers of local taxation, to be dealt with solely in the interest of the drainage of the country,

while the latter[19] reported as follows:

> it is quite clear on the evidence that the returns of the navigation have been ridiculously small, amounting on an average of the last twenty-three years to less than £70 per annum, of which sum a large proportion is due to the fishing boats entering the river by the Toome lock, and merely proceeding about a quarter of a mile to Toome Railway Station . . . looking to the fact that there is a railway on both sides of the River Bann, we have difficulty in believing that any traffic on the navigation of the Lower Bann will hereafter be realized and if it had not been for the action of the Grand Jury of County Derry in 1883, who decided on maintaining the navigation of the Lower Bann, we should have had no hesitation in recommending that the navigation works should be abandoned and the locks and lock cuts made available for the discharge of floods . . . if it were hereafter desirable to restore the navigation, this could be done at a moderate cost by refixing the lock gates and carrying out the river improvements which the use of the navigation works for the discharge of floods would for the time render unnecessary.

This Allport Commission of 1887 recommended that whether the navigation on the Lower Bann be retained or not, a single Conservancy Board should be formed for the whole catchment area of the Bann and, if the navigation were continued, should be entrusted with its maintenance and management.

There was little result from the recommendations of either Commission and the navigation remained open during the remainder of the century, despite decreasing traffic and receipts. It should, however, be borne in mind that expenditure on the Lower Bann Navigation included the cost of works, such as the dredging of Lough Beg, which were necessary for drainage as well as navigation, and had navigation been abandoned, the responsibilities of the Lough Neagh Drainage Trustees and the financial levy on riparian landowners for drainage works would have been increased. On the other hand, the use of the locks and lock cuts solely for the discharge of flood water would have meant a considerable saving in the cost of maintaining effective drainage throughout the Lough Neagh basin, a saving estimated in 1887 at some £10,000 per annum.

At any rate, the conflicting interests of drainage and navigation of the Lower Bann had not been resolved as the nineteenth century drew to its close: the failure of the navigation and the disappointing nature of the drainage works had not been remedied nor had any decisive measures been taken towards closing the former or improving the latter.

When Sir Alexander Binnie, President of the Institute of Civil Engineers, visited the area at the invitation of the Government in the late summer of 1905, commercial navigation on the river had almost ceased:

> during residence of over seven weeks at Brecart Lodge near Toomebridge I never saw any traffic of any kind, with the exception of a pleasure steamer on two occasions, passing up or down the navigation[20]

and after a comprehensive investigation into the conflicting claims and interests of drainage and navigation in the area he concluded

> that if the question of reducing the winter level of Lough Neagh is to be accomplished at any reasonable expenditure, it will become necessary to entirely abandon the navigation.

He regarded the basic concept of utilizing the Lower Bann as both a drainage and a navigation channel as ill-founded and incompatible with existing physical conditions—

> putting all other matters on one side and regarding it as a canal for economical traffic, it violates the first principles of canal engineering, for the whole economy of inland navigation is the maintenance of still water ponds between the different locks along which the navigation can be handled at a low cost. In the case of the Bann Navigation, however, we have a canalised river down which passes, against any upward traffic in the winter months, floods at the rate of 400,000 to 800,000 cubic feet per minute. It is therefore not surprising to me that the navigation has not proved a commercial success.

In conclusion Binnie pointed to the economic and social factors that made any increase in water transport unlikely:

> the whole area is well served by railways and looking at the country generally, I do not see that it has any chance of improving in the future. The railways may be said to entirely encircle Lough Neagh

and there are practically two lines of railways down the Lower Bann, bringing the whole district into railway communication with Larne, Belfast, Newry, Dublin, Coleraine, Londonderry and Portrush.[21]

Flooding in the lower regions of the Bann valley and round the shores of Lough Neagh remained so serious that before yet another inquiry in the following year,[22] this time a Royal Commission to inquire into all the waterways of the United Kingdom, a determined effort was made by landowners of areas subject to inundation to have the navigation finally abandoned and its works demolished. Opinions were again divided; there were still those who regarded the navigation as an essential link in internal communications in the north of Ireland and who did not agree that the maintenance of the navigation and efficient drainage were inherently incompatible. After weeks of protracted and painstaking examination the Commission recommended that the river should be maintained primarily for drainage, but that the navigation works should not be removed in case traffic should revive. Modifications in the weirs and more efficient dredging of the navigation channel might be necessary to facilitate drainage and prevent flooding and the Lower Bann Navigation Trustees should be left in no doubt as to their responsibilities in this direction.

Though the navigation works had once again survived, there was no great improvement in traffic or receipts on the river in subsequent years and most of the commercial trade was confined to the more southerly section of the river, in and around Toomebridge at the northern end of Lough Neagh. In addition, the river carried a little pleasure traffic. Though the tonnage handled at Coleraine had increased after the improvements of 1879–84, practically none of it moved upstream of Coleraine bridge by water. Goods arriving at the port for further distribution went by rail. Furthermore, at this time there was a greater traffic in coal, meal, potatoes, and grain from Portadown, Lurgan, Lisburn, and even Belfast to places like Toomebridge, Portglenone and Kilrea, than from Coleraine.

The Trustees continued to maintain the navigation works and channel at a substantial loss, the excess of maintenance cost above annual receipts being met by the ratepayers of certain areas in Counties Antrim and Londonderry, as laid down in the Award of 1859. The discrepancy between expenditure and receipts on the Lower Bann was of the order of £1,500 to £2,000 per annum, tolls and wharfage often yielding less than £100 and maintenance requiring over £1,500.

By the Drainage Act (N.I.), 1925,[23] and the Antrim Electricity Supply bill introduced at Stormont towards the end of the same year, navigation interests on the Lower Bann were severely threatened. By the former the Lower Bann Navigation Trustees were placed in some uncertainty as to their position and responsibilities. Though nominally a Navigation Trust, most of their work in the past had been in the interest of drainage, and it now seemed as if this work would henceforth become the responsibility of County Councils or other bodies.

The proposal to establish a dam near Carnroe lock[14] to pond back the river and create a reservoir from whose regulated outflow hydro-electricity could be generated and supplied to the greater part of County Antrim would have meant cutting the navigation into two separate sections, above and below the proposed barrage. Though this project never got beyond the planning stage, it contributed to the anxiety and frustration of the Trustees during the final years of their existence and helped to smother any chance which the navigation may have had of attracting traffic.

By the Act of 1925 a Drainage Advisory Committee was established to consider drainage problems on Lough Neagh and the Lower Bann and, having carefully examined both the area as it then existed and the prolonged and complex history of past attempts to alleviate flooding, it issued two reports, the first in 1927, the second in the following year.[25] In these, which mainly dealt with the disposal of surplus water from Lough Neagh, the main recommendation that concerned the navigation entailed the provision of sluices at Toome, Portna and the 'Cutts' to enable water to be discharged in sufficient volume to prevent flooding around the shores of the Lough and in the Upper Bann and Blackwater valleys and the execution of those works in the channel of the Lower Bann necessary to render it capable of carrying away the flood water discharged into it:

> as regards navigation north of Toome weir, care must be taken to avoid conflict of the various interests of drainage, power, fisheries and navigation in dry seasons . . . navigational interests will naturally wish to limit the outflow at Toome in summer so as to avoid the risk of creating shallows in the Lough and canal approaches, whereas power interests would be better served by a continuous withdrawal of water from the Lough . . . by providing the control of Lough Neagh from a level of 52·5 feet above Ordnance Datum it will be possible to provide sufficient water for the fisheries and for a restricted output of electrical energy without deleterious effects on the

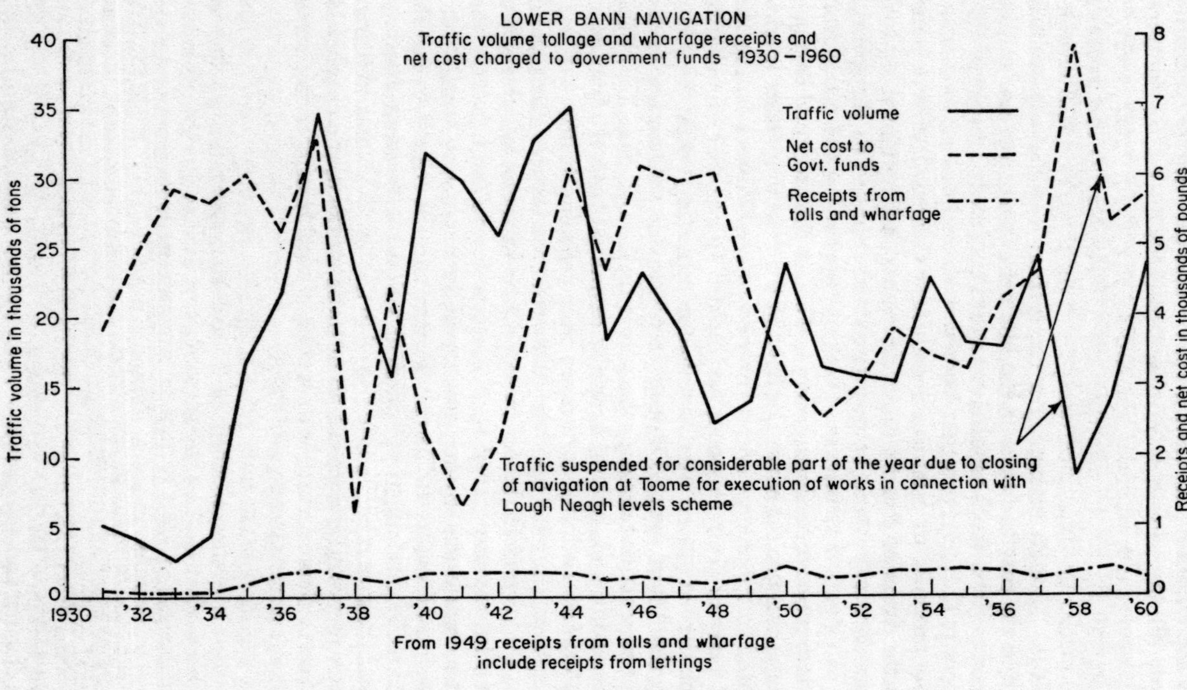

17. Lower Bann Navigation graph

navigation north of Toome weir . . . should a hydro electric dam be erected near Carnroe lock it is recommended that the Company should provide landing places adjacent to and above and below the new dam and should install approved facilities for the free transport of small boats and launches from one reach of the river to the other.[26]

As it was soon realized that the total cost of the scheme outlined in the interim report would be something over one million pounds, a sum likely to prove prohibitive, the final report of 1928 outlined a modified programme involving the provision of sluices at Toome weir, the improvement of the entire river channel from Toome down to Portna rock barrier and the execution of such minimum works there, including the removal of Portna weir and the substitution of sluices, as would prevent this stretch from rendering ineffective the improved channel upstream. These undertakings, the minimum required to reduce the winter level of Lough Neagh below flood level, were to cost £650,000, of which the County Councils were to provide £200,000. Finally, the Drainage Advisory Committee recommended that from the commencement of the works, the Ministry of Finance in the Government of Northern Ireland should take over the maintenance of the channel of Lower Bann, which would of necessity entail the abolition of the Lower Bann Navigation Trust.

As a result of these reports was framed the Drainage Act (N.I.), 1929,[27] by which the Lough Neagh Drainage Trust and the Lower Bann Navigation Trust were abolished and their powers and duties transferred to the Ministry of Finance, who were henceforth responsible for both drainage and navigation on the northern portion of Lough Neagh and on the Lower Bann.

Since the transfer, traffic on the Lower Bann has consisted mainly of building materials (stones, gravel, sand, cement, bricks, roofing felt, slates, etc.), raw materials (ores, cotton, wool, metal, leather, clay, etc.) and coal. The Government has been concerned with the navigation only from a 'works' point of view, in the adequacy of the channel of the Lower Bann to discharge surplus flood water rather than in its commercial success as a line of inland communication. It has not been particularly interested in the volume of traffic using the river and locks or in the continued failure of receipts to balance expenditure.

Since 1930, receipts from tolls and wharfage have only to a very slight extent offset expenditure on the maintenance of the navigation channel and associated works. It was fully realized at the time of transfer that the navigation would prove nothing but a liability and it has not been Government policy to spend large sums of money on major improvement schemes. The only direct contribution which the Ministry has made to the navigation was the construction of a dry dock at Portna in 1942, largely because of a temporary increase of traffic created by the wartime emergency. The Ministry has also from time to time carried out drainage works which have benefited navigation, for instance, the replacement of weirs by sluices (flood gates) at Toomebridge, Portna, and the 'Cutts' in the 1930s and the deepening of the Toome canal during the carrying out of the Lough Neagh (Levels) Scheme implemented between 1955 and 1959 under the Lough Neagh and Lower Bann Drainage and Navigation Act (N.I.), 1955, by which the level of Lough Neagh has to be maintained within the range of 50–50½ ft O.D., so long as weather conditions permit. On average between £8,000 and £10,000 is spent annually on the Lower Bann drainage conduit.

Today the bulk of the traffic on the river is contributed by three firms. Messrs. H. & W. Scott, sand merchants, Toomebridge, who succeeded the Lough Neagh Sand and Brick Company, use the navigation fairly frequently between Lough Neagh and Toome wharf with diesel barges, originally steam-powered, that were mostly built between 1929 and 1931 for Arthur Guinness, Son & Co. of Dublin, and also the United Kingdom Peat Moss Litter Company, Toomebridge and the Diatomite Company, Newferry, both of whom occasionally use the river between Portglenone and Toomebridge, mainly for local transport between their various workings on each side of the river. A few other sand and gravel merchants operate boats on Lough Neagh but do not use the navigation except when bringing a newly-acquired craft into the Lough, while ten or twelve motor launches, pleasure craft, canoes, etc., also use the navigation from time to time.

It is now a question of how long the navigation will be maintained by the Government, abandonment having already been discussed at some length in official circles.

Upper Bann Navigation

The Upper Bann Navigation extended along the river Blackwater from Lough Neagh to Blackwatertown and along the Upper Bann from Lough Neagh to Whitecoat Point, including also the southern portion of Lough Neagh. The length of the two sections of river channel included in the navigation was 21 miles 28 chains. There were no locks.

The improvements carried out between 1847 and 1858 and subsequently handed over to the Upper Bann Navigation Trustees included the following:[28]

(i) the Maghery Cut,* with bottom breadth widened to sixty feet and excavated to a depth of forty to fifty feet above datum;

(ii) fixed timber staging, twelve feet wide, extending partly across the said Cut, forming a roadway to meet Lord Charlemont's pontoon bridge;

(iii) the channel of the Blackwater river, from Maghery up to Blackwatertown, not including the embankments and sluices along the river, which are to be maintained by the Lough Neagh Drainage Trustees;

(iv) general breadth of river, from eighty to one hundred feet, and depth from Verner's Bridge to Moy to be five feet six inches in summer, reducing to five feet approaching Blackwatertown and a portion from Maghery to Verner's Bridge being considerably deeper;

(v) the improved channel through the bar in Lough Neagh at the mouth of the Upper Bann, ninety feet wide and five feet six inches deep at reduced summer level of the Lough which is forty-six feet above Ordnance Datum. The channel to be effectively marked with piles.

These were the works which it was the duty of the Trustees to keep in order and which they maintained with reasonable diligence in subsequent years. They were empowered to levy tolls and any deficit resulting from the maintenance and operation of the navigation by the Trustees was to be met by the ratepayers of certain areas in Counties Tyrone, Armagh and Down (see p. 129) in proportions fixed by the Award of 1859.[29]

In 1846 McMahon had not only envisaged a speedy repayment of capital by the counties but also an annual income from tolls and water power yielding a dividend of $5\frac{1}{4}$ per cent to all the areas included in the financial schedule of 1846. The bulk of this income would accrue from the Lower Bann Navigation but would be paid to *all* the areas assessed by Harding in 1846. Yet ten years later an Act[30] had been passed severing the navigation works and giving financial responsibility for those on the Lower Bann to Counties Antrim and Londonderry and those around Lough Neagh and on the Blackwater and Upper Bann to certain areas in

* This was a passage between Derrywarragh Island and the mainland. The pontoon was movable, so that craft could pass.

Counties Armagh, Tyrone, and Down, which would forfeit all right to the income accruing from tolls and mill power on the Lower Bann. As we have seen, thanks to railway development in the Lough Neagh basin, which in the interim had seriously blighted the prospects of the Lower Bann as a commercial waterway, the areas of Counties Armagh, Down and Tyrone for whom an income of over £300 a year had been envisaged by McMahon in 1846 were by 1856 perfectly willing to give up their chance of obtaining it.

In addition to the improvement outlined above, a quay was constructed at Moy on the Blackwater and an improved harbour on the shore of Lough Neagh at Kinnego, near Lurgan. Other work subsequently carried out under the auspices of the Upper Bann Navigation Trustees included a wharf at Blackwatertown, built as a result of a request by Armagh merchants who considered the railway rates to the town excessive and hoped that the establishment of this alternative route for heavy traffic would cause the railway company to reduce them. They were successful, for railway rates were reduced to such an extent that the facilities ceased to be used after about a year. However, the wharf was kept in repair by the Trustees as a deterrent against a subsequent increase. A slip and boat yard were built at Maghery where the various dredgers, barges and boats belonging to the Trustees could be stored. This also served as a repair yard, thus obviating the expense of transport to Belfast or Portadown for normal maintenance and minor repairs.

During the years immediately following its opening the Upper Bann Navigation was moderately successful, mainly by facilitating the movement of lighters up the Blackwater from Lough Neagh and the Lagan Navigation to the entrances of the Coalisland or Ulster Canals. Largely due to the failure of the Ulster Canal, a decline in the use of the Coalisland Canal and the attitude of the Lower Bann trustees, tolls and rents collected by the Upper Bann Navigation Trust were far from adequate to cover the cost of maintenance. Between 1864 and 1889 the total amount contributed by taxation was £22,591, whereas the tolls received each year never exceeded £100.[31]

During the period 1880 to 1905 there was an improvement in both tonnage carried and annual receipts, and for a time receipts exceeded expenditure. Indeed, during the twenty-five years, 1889 to 1914, the total amount contributed from the county rates decreased to £17,900, a drop of £4,691, compared with the preceding twenty-five years, and receipts increased by £3,118 over the same period.[32]

One cannot measure the value of the works solely by the traffic carried. It is safe to say that had it not been for the navigation the inhabitants of the district would have had to pay an increase in railway rates greater than the sum which they contributed towards the maintenance of the navigation works. The income from tolls, sales of washed sand and gravel dredged from various stretches of the Blackwater and Upper Bann, and rents is therefore not altogether a true criterion of the value of the navigational facilities. To cite but one example: in 1891, at the earnest request of farmers and merchants living in a wide area extending for some eight or ten miles between the Great Northern railway line from Dungannon to Cookstown and the lough shore, the harbour at Newport Trench, near Arboe, on the western side of Lough Neagh in County Tyrone was reopened and enlarged. Formerly a simple harbour or slip of some kind had existed at 'The Battery' or 'Duff's Battery', where vessels could unload goods from Lurgan and Portadown, the Lagan and Newry Canals, but the lowering of the level of Lough Neagh by the works of the 1840s and 1850s had made it useless. A new concrete harbour was constructed largely as a result of the exertions of J. B. Gunning-Moore, a prominent landowner in the area and one of the Trustees of the Upper Bann Navigation and trade soon developed during the early years of the present century.

The other main landing quays whose maintenance was the responsibility of the Trustees were at Ballyronan and Kinnego. The former was founded in 1788 by David Gaussen, of Huguenot descent, and a quay, spade factory, brewery, distillery, and corn mill established soon afterwards. By 1837 vessels of fifty tons were plying regularly to Belfast and Newry, exporting wheat, fruit, spirits, ale and freestone and bringing in barley, timber, slates, iron, wine, and groceries. By the works carried out between 1847 and 1858 the importance of the landing place there declined with the decreased draught available and although subsequent dredging was carried out, often by the Lower Bann Navigation Trust, and a grain import maintained until after the 1914–18 War, the introduction of commercial road transport did much to offset the five-mile haul to the nearest railway, at Magherafelt or Toomebridge, and traffic soon disappeared in the inter-war years.

Kinnego, or Port Lurgan, as it was sometimes called, was a navigational cul-de-sac at the south-eastern corner of Lough

Neagh, some two and a half miles from Lurgan. The harbour here handled a sizeable passenger traffic from Ballyronan and Newport Trench *en route* to fairs and markets in County Armagh or to the Ulster Railway at Lurgan, while large quantities of heavy goods, especially coal, were imported for the linen mills of Lurgan from the Lagan and Newry Canals during the latter half of the nine-teenth century.

At Ellis' Gut, the approaches to which were the responsibility of the Upper Bann Navigation Trust, lighters entering Lough Neagh from the Lagan Navigation paid tolls to representatives of the Trustees before being towed across the lough in tug trains to Portadown or to the entrance of the Blackwater and the bulk of their income was collected at this point.

Again, the maintenance of the Blackwater channel and estuary in a navigable condition was of considerable assistance in the draining of a large area south of Lough Neagh, for had the mouths of this river and of the Upper Bann been allowed to silt up, efficient drainage would have been impossible and serious flooding would have occurred in the catchment areas of both rivers.[33]

In spite of the financial improvement that we have noted between 1880 and 1905, and the consequent lessened calls upon the ratepayers, the continued demands on certain areas of County Tyrone which derived little or no benefit from the navigation works caused considerable discontent during the early years of this century and a comprehensive investigation into the whole structure of the Upper Bann Navigation rate was carried out by an independent observer.[34] From this inquiry it emerged that although the Upper Bann Navigation Trustees had been empowered to levy tolls up to 2s 6d per ton for any one voyage on all merchandise using the navigation, the tolls at the time of World War I were roughly $\frac{1}{2}$d per ton from which an annual revenue of some £200 was realized. Had the navigation rate in operation been fixed at 2d or even 3d per ton, a not unreasonable figure when comparison is made with tariffs in operation on other inland waterways in the north of Ireland at this time,* not only would there have been

---

* On the Lagan, Ulster and Coalisland Canals, for example, rates of toll (per ton mile) worked out as follows:

    (a) Lagan   (i) inwards, between $1\frac{3}{8}$d and $2\frac{1}{2}$d.
                         (ii) outwards, between $\frac{1}{3}$d and $\frac{1}{2}$d.
    (b) Ulster   (i) inwards and outwards, between $\frac{1}{3}$d and $1\frac{1}{2}$d.
    (c) Coalisland   (i) inwards, between $2\frac{1}{4}$d and 3d.
                         (ii) outwards, between 1d and 2d.

no annual calls on county revenue, but the substantial payments made from it to the Navigation Trustees in nearly every year since 1860 would have been available for distribution elsewhere.

Tolls were in fact increased in 1916 and again in 1920, from ½d per ton to approximately 3d per ton on all merchandise. Tollage receipts then showed a marked increase, but so did expenditure, and a county assessment of £300 per annum was still required. During this period the Trustees had complaints from many sources, the most repetitive being from the Lagan Navigation Company and lightermen regarding the unsatisfactory state of the channels leading to Newport Trench, Kinnego, Ballyronan and the entrance to the Coalisland Canal, from masters of vessels passing through the Maghery Cut about the obstruction and delay caused by the movable pontoon being left across the 'Cut' instead of being returned to its berth after use,[35] from John Gaussen of Ballyronan regarding the condition of the quay there, from the Lagan Navigation Company and Messrs Stevenson of Coalisland concerning the high tolls they considered they were being charged, and from Tyrone County Council with regard to the county assessment.[36]

By the beginning of the 1930s, indeed, the end of the navigation as a commercial undertaking was in sight and deputations appeared before both the Ministry of Commerce and the Ministry of Finance seeking either substantial financial assistance or a transfer of the navigation works and channels, and the abolition of the Upper Bann Navigation Trustees. As neither proposal was at that time acceptable to the Government, the Trustees were compelled to carry on as best they could, but despite stringent economy in both operational expenditure and maintenance, by 1936 the county assessment had risen to £800.

A further alteration of tolls in 1940 had little effect on annual receipts, which remained far below annual expenditure during the remainder of the navigation's existence. In 1941, for example, it was estimated that expenditure for the ensuing twelve months would be over £1,000, made up as appears on page 146[37]. Against this, anticipated receipts from tolls were estimated at £175 and miscellaneous receipts at £65, a total of £240, leaving a deficit of £807 and necessitating a county assessment of £800.

Kinnego Quay was taken over by the military in October 1940, being required as a base for patrol boats operating on Lough Neagh and a fixed annual rent of £50 was subsequently agreed on between the War Office and the Trustees. Once this payment ceased, however, receipts almost disappeared and after 1947 no traffic reached Kinnego, Portadown, Coalisland, Newport Trench, or Ballyronan from the Lagan Navigation. Peat moss and mould from the bogs of north Armagh formed the only commodity of importance leaving the Upper Bann and Blackwater for the Lagan or the Lower Bann.

	£
General Maintenance	250
Dredging:	
Coalisland Canal foot	90
Bannfoot	180
Blackwater fords	100
Coal, oil, etc.	40
Towage	15
Rent of sand depots and watchmen's wages	32
Salaries	340
	£1,047

As the navigation was still legally in existence and costing the ratepayers £800 a year, the Government at last decided to close the various navigation works and channels controlled by the Upper Bann Navigation Trustees, and to dissolve that body. By the passage of the Inland Navigation Bill (N.I.) of April 1954, the Tyrone Navigation, the upper Lagan Navigation and the Upper Bann Navigation were formally abandoned. Since then the channels of the Blackwater and Upper Bann have been dredged solely in the interest of drainage, and the various quays and harbours on these rivers and around the shores of Lough Neagh are no longer maintained for commercial traffic.

Thus, the navigation works and channels on the Blackwater and Upper Bann and around the shores of Lough Neagh, constructed between 1847 and 1858 and subsequently entrusted to the Upper Bann Navigation Trustees, had a relatively short life of rather less than a century. They were never a commercial success, their main practical use being to facilitate drainage in low-lying areas south of Lough Neagh and to impose a temporary curb on railway rates during a transitory phase in the development of communications in the Lough Neagh basin. For the greater part of their existence they constituted a steady drain on the finances of the neighbouring counties, and it was with no great difficulty that the bill providing for their final abandonment as a commercial undertaking passed into law in 1954.

CHAPTER VIII

Other Projects

‧‧‧‧‧‧‧‧‧‧‧‧‧‧‧‧‧‧‧‧‧‧‧‧‧‧‧‧‧‧‧‧‧‧‧◆‧‧‧‧‧‧‧‧‧‧‧‧‧‧‧‧‧‧‧‧‧‧‧‧‧‧‧‧‧‧‧‧‧‧‧

SCATTERED through the history of inland navigation in the north of Ireland are many plans for canals and river improvements that came to nothing. Six of these are worth mention.

1. Larne to Lough Neagh

In 1808 John Rennie was invited by James Agnew Farrell of Larne, who had just been granted a lease of some 290 acres in Larne Lough from the Marquis of Donegall, to carry out a survey of the land between Larne and Lough Neagh with a view to establishing a navigable waterway between these two points. The invitation was accepted and the survey carried out in 1808 and 1809.

Recommending various works of reclamation and pier construction in Larne Harbour, Rennie proposed a canal from either the mouth of the Larne River or the Bank Quays to cross the Larne River upstream by an aqueduct and thence trend northwestwards by Antiville and Kilwaughter to a summit level between Ballygowan and Ballynure. From the summit level the canal would descend to the Ballynure Water, which it would cross little more than a quarter of a mile below the town, where there would be a convenient place for a basin or wharf. From here it would pass along the south side of the Sixmilewater to Ballyclare paper mill, close to the southern edge of that town, where also a convenient basin or wharf might be made. The line would then cross the stream at Ballyclare and skirt the south side of the Sixmilewater near Templepatrick, finally entering Lough Neagh at the 'Black Rocks', about a mile south of Antrim town, where a pier would be built.

The total distance was about 18 miles and the estimated expense of making a canal 21 ft at bottom and 41 ft at top, with a uniform depth of 5 ft, was £168,137.

As an alternative, Rennie made an interesting proposal for the section from the summit level down to Larne Lough. He suggested an extension of the summit level by some 3½ miles to the road near the 'Four Towns' of Inver, a descent of 114 ft by locks to a basin and, finally, an inclined plane, some 272 ft in length, connecting with the pier at the Bank Quays, down which goods might be sent in waggons to load the sea-going vessels. The plane would be counterbalanced, the descending heavy waggons bringing up the lighter. This method was then in operation in the limestone quarries at Carnlough and was later used with success by Howdens Ltd., coal importers, for the transport of stone on or about the site which Rennie had proposed. It was also suggested that the waggon elevating gear could be worked by surplus water from the canal but the details of this proposal are not given. The cost of this scheme was estimated at £139,892.

Water from Rennie's canal was to come from the Larne and Sixmilewater rivers, both of which rise in high ground through which the cut would pass, supplemented by two reservoirs, one at Castletown, near Ballynure, with an area of 95 acres, the other at Ballyfore Moss, with an area of 35 acres.

The optimistic Farrell further suggested that the canal could be extended from Antrim to Randalstown and Ballymena at a trifling expense. A Government grant was anticipated, but in the meantime it was proposed to issue transferable shares of £50 or £100 each.

John Rennie was an authority on canal construction but not on County Antrim farmers. The required capital for the scheme was not forthcoming and James Agnew Farrell emerged a sadder and wiser man, poorer by some £800 paid as survey fees and preliminary expenses on a project which never advanced beyond the planning stage.

2. Coleraine to Portrush

In 1814 G. V. Sampson wrote: 'several projects for improving the harbour of Coleraine had been suggested. A canal to pass from the dam behind the town to the road of Portrush. A rail-way to the same place to accompany the new line of road. A canal from the river near the Laughin island to the above mentioned basin or dam; all of these have been spoken of but none of them has hitherto been much pressed upon the public attention not because the practical ability is doubtful, but because the advantages uncertain and the risque of course would be serious.'[1]

This proposed canal was felt to be needed because of the dangerous navigational approach to the port of Coleraine through the 'Bar-mouth' and the lower estuary of the Bann. But when a major harbour was built at Portrush under John Rennie's direction between 1827 and 1836, initially planned as an out-port for Coleraine but soon enabling Portrush to become a major port in its own right, the need for such a navigation disappeared. With the coming of the railways in the late 1840s and early 1850s the proposal was never revived.

3. Lough Foyle to Lough Swilly

The idea of cutting a canal across the low-lying isthmus, six miles wide, that separated Londonderry, on Lough Foyle, from Lough Swilly was kept alive over a long period. It was brought before the Irish House of Commons in 1763 and 1765, when a grant of £8,000 was unsuccessfully sought. In 1807 Londonderry Corporation had a survey made, but without result. In 1831 John Rennie made surveys and prepared three alternative plans at a cost of about £38,000 for each.

The project became involved with others to control the Swilly tides by embankments, and to reclaim mudflats. An Act of 1838 authorized these two projects and recognized that the canal might thereafter be built. However, the technical difficulties put paid to any such scheme.[2]

4. Strabane to Lough Erne

There is reference to this canal in 1811 with the appearance of a 'Prospectus of a plan for making a navigable canal to be called the North West of Ireland Canal'.[3]

This canal is intended to connect Lough Foyle with Lough Erne. There are two lines by which this object can be effected; one by commencing where the Strabane Canal ends and passing the river Foyle by an aqueduct; the other by commencing on the Donegal side of the Foyle, opposite to the Strabane Canal, passing through or by Lifford, Strabane, Newtownstewart and Omagh into Lough Erne. . . .

The prospectus, after outlining the great advantages which will result from such a link, concludes:

> It may be well to observe that one of the causes of failure of profit in Irish canals has been the splendid scale on which they have been made: this, therefore, will be executed on that business like, economical and profitable scale of those in the central counties of England.

The proposal recurs in 1824[4] and in 1834. Lieut. William Lancey of the Ordnance Survey wrote: 'The late Lord Blessington wrote a pamphlet to show the practicability of running a canal from Strabane to Lough Erne which would pass along the banks of the Mourne and Strewal [now Strule] rivers through the forest. The expense deterred from the undertaking those interested about it. A rail road would be a far more preferable method of rapid and easy communication and as stone is so abundant would probably cost far less money. The traffic between Omagh and the canal head at Strabane is in certain seasons considerable but it is much to be doubted if any undertaking of the above nature would pay its expenses where food and labour are so cheap and money so scarce.'[5]

5. Armagh to the Blackwater

In the Commons Journal of 23 June 1800 occurs the following report on a proposal to link the city of Armagh with the river Blackwater, which was probably part of a wider plan, being discussed at the same time, for a continuous line of inland navigation from Dublin to Lough Neagh, via Navan, Kells, Bailieborough and Monaghan:

> from Armagh, being the most considerable market in the Kingdom for the sale of brown linens, the manufacture of that staple article is carried on to a very great extent in its neighbourhood; but this manufacture is in danger of being most materially injured from the great scarcity of fuel, which is such as to oblige the opulent inhabitants to use English coal at a great expense of land carriage; and they have latterly, at inclement seasons, been under the necessity of subscribing large sums to procure that article at a low price for the poor, to prevent them from perishing.
>
> Should a Navigation be opened to Lough Neagh it would give the means of supply of turf from the extensive bogs in the neighbourhood of the lake, would open a communication with the Collieries at Coal Island in the County of Tyrone and bring English and Scotch coal considerably under the prices at which they can now be procured

. . . to the great number of Bleach Greens and Flour Mills in the neighbourhood of Armagh, water carriage would be of the highest importance, as well for the conveyance of Bleaching Stuffs, Coals, Grain and Flour, as of Timber, Slates and other heavy articles used in erecting and repairing the necessary houses, machinery, etc.

In a large tract of country from Blackwatertown to Lough Neagh and from thence up the River Bann and along the canal to Newry, there is no limestone whatever . . . was a canal opened to Lough Neagh from Armagh through the Blackwater it must go through lands containing large quantities of limestone which could be conveyed by boats returning from Armagh.

In 1815 John Killaly included a navigation offshoot to Armagh in his proposed line for the Ulster Canal, while in 1850 the area was surveyed by John Godwin and James Boyle with a view to constructing a railway from Armagh to Caledon (see Ch. VI, note 11).

Despite these various proposals neither canal nor railway was ever built to link Armagh with the Blackwater or the Ulster Canal, though the Ulster Railway extension to Monaghan, completed in 1858, was carried over both river and canal near Tynan.

6. Limavady to Lough Foyle

In 1827, at the invitation of a number of farmers, merchants, and landed proprietors in the Roe valley, in north Londonderry, John Killaly, one of the engineers attached to the office of the Directors General of Inland Navigation in Dublin, made a survey of the area between Limavady and Lough Foyle with a view to constructing a navigable canal. Killaly's estimate for the proposed cut was £12,155 and he was of opinion that from a harbour or basin in the townland of Shanreagh, about a mile west of Limavady, to Ballymacran Point, the canal would have a total length of 3 miles 10 chains with two locks, one at the confluence with the Foyle, the other at a point about a mile from the inland terminus. The canal was to be 20 ft wide at bottom, 35 ft wide at top with 5 ft depth of water: the locks were to be 60 ft long 'within the gates' and 13 ft 6 in. wide at the top. It was anticipated that such a canal would soon handle a substantial outward traffic in agricultural produce, butter, pork, beef, and bricks with timber, slates, shells (for manure) 'kelp' or seaweed (another important import used as fertilizer), and above all coal the principal imports.

Nothing came of this proposal and in 1832 we find a meeting of 'landed proprietors, merchants and inhabitants of Newtown-limavady' calling for the construction of 'a projected rail road from Newtown Limavady to Lough Foyle, a distance of three English miles, upon an inclined plain of about seven feet to each mile from the commencement of the line at the Roe Bridge to the termination thereof at Ballymacran'. A similar traffic to that envisaged for the proposed canal was anticipated for this horse rail-road, the construction of which was to be coupled with that of a pier at Ballymacran to enable the barges operating on Lough Foyle to discharge coal and timber or to take on grain, meat, pork or bricks at all stages of the tide.

Author's Notes
and Acknowledgements

A STUDY of the emergence and decline of an important transport system involves a great deal of documentary research in order to create an adequate picture in breadth and depth. Amongst the very many people to whom I am deeply indebted for helpful guidance and patient co-operation during the writing of this book I owe particular thanks to the staffs of the several libraries which I used—the Queen's University Library, the Linen Hall Library and the Central Public Library in Belfast, the library of Trinity College, the Royal Irish Academy and the National Library in Dublin; to Miss Margaret Griffith of the Public Record Office Ireland; to Mr K. Darwin, Mr B. I. Trainor and Mr B. G. Hutton of the Public Record Office Northern Ireland; to Mr W. J. Forster of the Ministry of Commerce in the Government of Northern Ireland and to Mr L. M. Buchanan of the Ministry of Finance; to Mr B. O'Rorke of the Newry Port and Harbour Trust; to Mr T. G. F. Paterson and Mr D. R. M. Weatherup of the County Museum, Armagh; to Mr W. J. Guckian of the Office of Public Works in Dublin; to the Knight of Glin and to the late Prof. V. T. H. Delany.

Many of the maps have been prepared by the cartographic staff of the Department of Geography at the Queen's University of Belfast and I owe a particular debt of gratitude to Miss Eileen Duncan, and to her assistants, not only for the skilful drawing of many of the maps and diagrams but also for much helpful discussion, particularly in the cartographic handling of statistics. Similarly, Mr T. J. Maguire and Mr L. Salem have done much to contribute to the large number of illustrations which I have been able to include.

In conclusion, to Professor E. Estyn Evans of the Queen's University of Belfast and to Dr E. R. R. Green of Manchester University, from whom I received much helpful, constructive criticism, astute guidance and stimulating interest; to the many country folk I met during the course of my travels about the province, and from whom I invariably received both hospitality and assistance; to my publisher, Mr Charles Hadfield, who has been most helpful in the preparation and editing of a great mass of material; and lastly, to my wife and secretary, without whose typing skill and general co-operation progress would have been greatly retarded, I take this opportunity of expressing my sincere thanks.

Thanks are due to the following for permission to reproduce photographs: Survey of Industrial Archaeology, Ministry of Finance, Northern Ireland, 1, 2, 3, 4, 6, 7, 13, 14, 15, 16, 19, 20, 22, 23, 26, 28, 29, 30, 31, 33, 34, 35, 37, 38, 39; *Belfast Telegraph*, 5, 32; Newry Port & Harbour Trust, 8; Welch collection, Ulster Museum, 9; Dr E. M. Patterson, 10; *Belfast News-Letter*, 11, 12; Works Division Engineers, Ministry of Finance, Northern Ireland, 17, 18, 36; Public Record Office, Northern Ireland, 20, 24, 25; Armagh County Museum, Fig. 4; Eugene Gormley, Strabane, 27.

NOTES

Notes to Chapter II (p. 17)

1. Legend appearing on Francis Nevil's survey of 1703, Public Record Office Northern Ireland: D.695/M.
2. Thomas Ashe, *A View of the Archbishopric of Armagh, made and taken in 1703*, Dublin, 1703.

   A portion of this appears in the *Newry Commercial Telegraph* of 6 June 1818, P.R.O.N.I.: T.848, p. 35. See also *Journals of the House of Commons of the Parliaments held in Ireland* (J.H.C.I.), 30 September 1703, III, p. 25; 20 May 1709, III, p. 426; 28 May 1709, III, p. 441; Brewster's *Edinburgh Encyclopaedia* 1830, p. 267.
3. 2 Geo. I (1715). This was the first Statute relating specifically to inland navigation in Ireland.
4. 3 Geo. II (1729).
5. MS. 'Essay on Artificial Navigation' by Richard Castle, offered at the Carton sale in June 1949 and now in the National Library of Ireland (MS. 2737). This reveals Castle's sound working knowledge of canal engineering and basic mathematical physics.
6. Report 13 of the Hist. MSS. Comm., 1893, App. 6, p. 197 (Delaval MSS.); *Faulkner's Dublin Journal*, 27 February–3 March 1732.
7. *Faulkner's Dublin Journal*, 4–7 May 1734.
8. J.H.C.I., 7 May 1768, VIII, Pt. II, p. 249.
9. Ibid., and T. C. Barker, 'The beginnings of the Canal Age in the British Isles', pp. 6 *et seq.*, in ed. L. S. Pressnell, *Studies in the Industrial Revolution—Essays presented to T. S. Ashton.*
10. Henry Peet, 'Thomas Steers, the engineer of Liverpool's first dock—a memoir', *Trans. Hist. Soc. Lancs & Cheshire*, LXXII (1930), pp. 163–242.
11. *Dublin Evening Post*, 23 August 1737
12. *Dublin Evening Post*, 15–19 November 1737; 15–18 July 1738; 28–31 October 1738; *Faulkner's Dublin Journal*, 29 October–1 November 1737. See also J.H.C.I., 10 November 1737, VI, p. 746; 15 November 1737, VI, p. 760; 16 November 1737, VI, p. 761.
13. *Faulkner's Dublin Journal*, 18–22 August 1741.
14. *Dublin News-Letter*, 30 March 1742.
15. For a detailed description of the canal immediately on completion see W. Harris, *The Antient and Present State of the County of Down*, Dublin, 1744, pp. 115 *et seq.*
16. R. Barton, *A Dialogue of some things of importance to Ireland, particularly to the County of Ardmagh*, Dublin, 1751, p. 146.
17. Ibid., p. 18.
18. J.H.C.I., 10 November 1767, XIV, p. 375.
19. J.H.C.I., 14 November 1749, VIII, pp. 111–12. See also M. B. Mullins, 'An Historical Sketch of Engineering in Ireland', in *Trans. Inst. Civil Engineers of Ireland*, VI (1859–61), pp. 1–181.
20. J.H.C.I., 31 March 1750, VIII, pp. 231–8.

21. J.H.C.I., 1 November 1755, IX, p. 328; 8 November 1755, IX, p. 376; 12 November 1755, IX, pp. 385–6.

22. J.H.C.I., 14 November 1759, XI, p. 243.

23. Two maps in the P.R.O.N.I. show the nature of the canal extension then in progress, 'A Plan of the Town of Newry' by Matthew Wren (1761), T.618/327, and 'A Survey of the Town of Newry' by Bernard Scale (1763), D.671/M9/5.

24. Considerable sums were laid out between 1760 and 1767 without producing any result and in the latter year a Committee of the House of Commons, inquiring into progress on the new navigation works, reported them to be wholly inoperative (J.H.C.I., 10 November 1767, XIV, p. 375). Appearing before this Committee, Christopher Myers, architect to H.M. Board of Works, stated that so much of the work as had been carried out by him between the time of Golborne's departure in 1760, and 1763, was in accordance with a plan approved by the House in 1759. He also observed that he had offered to complete the works for £18,000, provided half the amount was granted at once and the remainder in the following session. As only £5,000 had been immediately forthcoming he refused to accept responsibility for the delay in completing the scheme.

25. Gordon's *Newry Chronicle and Universal Advertiser*, 13 March 1785.

26. Arthur Young, *Tour in Ireland in 1777 and 1778*, 2 vols, 1780, edited by A. W. Hutton and reissued in 2 vols in 1892, I, p. 116. When Young visited the town in 1777 he was deeply impressed by the new ship canal which he described as 'indeed a noble work with ships of 150 tons, and more, lying on it like barges on an English canal'.

27. J.H.C.I., 8 May 1786, XXIII, pp. 353–6; 2 February 1788, XXV, pp. 62–3; 11 March 1789, XXVI, p. 139.

28. J.H.C.I., 23 June 1800, App: Vol. II, p. miv.

29. Gordon's *Newry Chronicle and Universal Advertiser*, 27 December 1792, 30 April 1793.

30. Public Record Office Ireland, Report Book of the Directors General of Inland Navigation, P.W. I. 5/7/1, pp. 26–44. (For full details of these years the MS. records of the Directors General of Inland Navigation—Letter Books, Report Books, Contract Books, Account Books, etc.—should be consulted in the P.R.O.I. at the Four Courts in Dublin.)

31. Report Book of the Directors General of Inland Navigation, P.R.O.I., P.W. I. 5/7/1, pp. 321 *et seq.*

32. *Thirteenth Report of the Commissioners appointed to inquire into the fees, etc., in certain public offices in Ireland: Directors General of Inland Navigation*, H.C. 1812–13 (61), VI, 317, pp. 17 and 53.

33. Sir Charles Coote, *Statistical Survey of the County of Armagh . . . drawn up in the years 1802 and 1803 for the consideration and under the direction of the Dublin Society*, Dublin, 1804, p. 108.

34. *Report made by order of the Board of Inland Navigation in Dublin, respecting the Newry Navigation*. H.C. 1826–7 (255), XX, 109, pp. 109–111.

35. 10 Geo. IV *c.* 126 (local and personal).

36. Sir John Rennie, 'On the improvement of the navigation of the River Newry', in *Min. Proc. Inst. Civil Engineers*, X (1850–1), pp. 277–93.

37. *Tidal Harbours Commission (Second Report)*: Appendix B, No. 131, H.C. 1846 (692), XVIII, Pt. 1, 1, pp. 224–9. For progress on engineering work during the period 1830–50, see the records of the former Newry Navigation Company, now in P.R.O.N.I. (D.934): Letter Books, Nos 1, 3 and 4: Minute Books Nos 1, 2 and 3.

38. Statistics from the Account Books of the Directors General of Inland Navigation, P.R.O.I., P.W. 1 5/9/3–8.

39. *Second Report of the Commissioners appointed to consider and recommend a general system of Railways for Ireland*, H.C. 1837–8 (145) XXXV, 449, Appendix B, No. 6, p. 59.

40. Statistics from *A Handbook to Carlingford Bay*, London, 1846. For further information on the navigations at Newry in the 1830s and 1840s see Samuel Lewis, *A Topographical Dictionary of Ireland*, 2 vols, London, 1837, *II*, p. 431; *The Parliamentary Gazetteer to Ireland*, 3 vols, Dublin and London, 1844–6, *III*, pp. 23 *et seq.*

41. Statistics from Newry Navigation Company records in P.R.O.N.I.: D.934/4, Letter Book, 1848–51, D.934/7, Letter Book, 1866–76, D.934/19, Minute Book of Annual General Meetings, 1830–1901.

42. Anthony Marmion, *The Ancient and Modern History of the Maritime Ports of Ireland*, London, 1855, pp. 303–17.

43. *Final Report of the Royal Commission appointed to inquire into the Canals and Inland Navigations of the United Kingdom*, H.C. 1911 [Cd. 5626] XIII, 19: pp. 30 *et seq.*

44. Portlock, in his valuable Geological Report of 1843, estimates that between 1829 and 1842 some 219,000 tons of coal were sold at the various collieries belonging to the Hibernian Mining Company alone; of this only about one-sixth was exported by canal.

45. E. T. Hardman, 'Explanatory Memoir' to accompany Sheet 35 of the maps of the Geological Survey of Ireland. Dublin, 1877, pp. 39–66.

46. Marmion, *Maritime Ports*, pp. 311, 353, 413.

47. *Report of the Commissioners appointed to inquire respecting the system of Navigation which connects Coleraine, Belfast and Limerick*, H.C. 1882 [C. 3173] XXI, 101.

48. Royal Commission, H.C. 1911 [Cd. 5626] XIII, 19: pp. 30 *et seq.*

49. Royal Commission, H.C. 1882 [C. 3173] XXI, 101: minutes of evidence nos 775–811.

50. Statistics from the annual *Reports and Statements of Accounts* of the Trustees of the Newry Port and Harbour Trust, kindly lent by B. O'Rorke, Secretary.

Notes to Chapter III (p. 40)

1. ed. R. P. Mahaffy, *Calendar of State Papers relating to Ireland*, 1633–47, London, 1901, p. 174. In 1637 Sir George Rawdon wrote to Lord Conway that he had 'made a proposition to Mr. Arthur Hill, who now owns Kilwarlin, to cut the Lough into the Lagan and make it portable to Belfast'.

2. J.H.C.I., 24 October 1753, IX, pp. 62–4; 1 November 1753, IX, p. 77; 6 November 1753, IX, p. 102; 8 November 1753, IX, p. 138.

3. 27 Geo. II (Irish) *c*. 3.

4. J.H.C.I., 14 November 1749, VIII, pp. 111–12; 31 March 1750, VIII, pp. 231–8.

5. Sec. I of 27 Geo. II (Irish) *c*. 3.

6. *Belfast News-letter*, 8 November 1757.

7. 33 Geo. II (Irish) *c*. 1 and 1 Geo. III (Irish) *c*. 1. See also J.H.C.L., 14 November 1759, XI, p. 238.

8. *Belfast News-letter*, 9 September 1763.

9. J.H.C.I., 18 November 1767, XIV, p. 375; 7 May 1768, XIV, pp. 541–3. By Section IX of the Act of 1753 the 'Corporation for promoting and carrying on an Inland Navigation in Ireland' had been empowered 'to divert the water of the said river Lagan from the said navigation in order to supply with water such bleach yards as are now watered by the said river, in dry seasons, where there would not otherwise be a sufficient supply of water to ensure the purposes of bleaching and of carrying on the said navigation'.

10. *A Plan and Estimate of the intended navigation from Lough Neagh to Belfast* as surveyed by Mr Robert Whitworth, together with his report concerning the best method of executing the work. Presented to the local committee at Hillsborough, August 24, 1768 and approved by Mr James Brindley, Engineer, Belfast, 1770.

11. 11 and 12 Geo. III (Irish) *c.* 26 and 13 and 14 Geo. III (Irish) *c.* 12.

12. 19 and 20 Geo. III (Irish) *c.* 32.

13. Letter Book, Lagan Navigation, 1810–14. P.R.O.N.I., C.O.M. I/22, p. 13.

14. *Belfast News-letter*, 31 December 1782.

15. It was at first proposed that the little bay and harbour at which the canal entered Lough Neagh should be called Port Chichester but this was never adopted and Ellis' Gut was the name given to the inland terminus of the waterway.

16. *Northern Star*, 28 March 1792; 31 March 1792. *Belfast News-letter*, 31 December 1793.

17. *Belfast News-letter*, 7, 11 February, 25, 29 August, 31 December 1794.

18. Minute Book, Lagan Navigation, 1793–1811. P.R.O.N.I., C.O.M. I/11, pp. 22–3, 27–9 and 40.

19. J. Dubourdieu, *Statistical Survey of the County of Down*, Dublin, 1802, pp. 26, 27.

20. *Belfast News-letter*, 6 March 1801.

21. Letter Book, Lagan Navigation, 1809–26. P.R.O.N.I., C.O.M. I/21, pp. 1, 7–10. Minute Book, Lagan Navigation, 1793–1811, P.R.O.N.I., C.O.M. I/11, pp. 68, 97 (this minute book also contains a 'Memorial to the Directors General of Inland Navigation', dated 28 May 1811: page unnumbered).

22. The Lagan Navigation always remained outside the control of the Directors General of Inland Navigation. They were involved on this occasion because of the request for financial assistance in the scheme of overhaul and improvement then being proposed.

23. For full information on the various schemes of extension and improvement being mooted at this time, see the records of the Directors General of Inland Navigation at the Four Courts in Dublin, e.g. (a) P.W. I 5/3/10 (Letter Book: Northern District) unnumbered page, a letter of 13 December 1811 from the Navigation Board to the Lagan Navigation. (b) P.W. I 5/3/10 (Letter Book: Northern District) unnumbered page, a letter from the Navigation Board (3 January 1812) to Lagan Navigation interests. (c) P.W. I 5/3/10 (Letter Book: Northern District), unnumbered page, a letter of 1 February 1812 from the Navigation Board to Joseph Stephenson of Belfast. (d) P.W. I 5/3/11 (Letter Book: Northern District), pp. 128–30, a letter from the Navigation Board (24 December 1813) to the Lagan Navigation.

24. J. Dubourdieu, *Statistical Survey of the County of Antrim*, Dublin, 1812, pp. 365 *et seq.*

25. 54 Geo. III *c.* 231 (local and personal).

26. *Belfast Monthly Magazine*, June 1810, IV, p. 477. Thomas Bradshaw's *General and Commercial Directory for the town of Belfast*, 1819, pp. xl–xlii. Bradshaw gives the following figures:

Year	Receipts	Expenditure	No. of Boats	Total Tonnage
1815	£2,299	£2,999	41	1,640
1816	£2,898	£2,505	44	1,760
1817	£3,997	£4,734	57	2,280
1818	£4,703	£4,423	68	2,720

27. *Belfast News-letter*, 8 January 1813.

28. With the continuing acute shortage of water on the summit level, the Lagan company again seriously considered tapping the Ravarnet and its head-streams to feed the canal. Once again, however, they encountered fierce opposition from the owners of the bleach greens along the river, particularly the Coulsons of Lisburn, who persuaded the Marquis of Hertford to refuse permission for any such works on his estates.

29. Letter Book, Lagan Navigation, 1809–26, P.R.O.N.I., C.O.M. I/21, pp. 198–9, 209–10, 243–4, 261, 269. Richard Owen died in 1830 and was buried in Soldierstown graveyard at Aghalee parish church. The inscription on his grave reads: 'Here lyeth the body of Richard Owen of Flixton, Lancashire, Engineer of the Lagan Navigation, who departed this life—1830.'

30. *Second Report of the Irish Railway Commission* (1838), op. cit., App. B, Sect. 6: Inland Navigations, pp. 50 *et seq.*

31. Lewis, *Topographical Dictionary of Ireland*, I, pp. 16, 279.

32. Manuscript notebooks compiled during the course of the first Ordnance Survey of Ireland, during the 1830s. The relevant parishes for information on the Lagan Navigation are Aghalee and Blaris, both in Co. Antrim (Boxes I and II in the Royal Irish Academy).

33. *Second Report of the Irish Railway Commission* (1838), op. cit., App. B, Sect. 6: Inland Navigations, pp. 50 *et seq.*

34. Letter Book, Lagan Navigation, 1809–26, P.R.O.N.I., C.O.M. I/21, p. 182.

35. *Report of the Select Committee on the Lagan Navigation Bill*, H.C. 1842 (537) XIV, 417.

36. 6 and 7 Vic. *c.* 104 (local and personal).

37. There was never any passenger traffic on the Lagan Navigation, so that competition in this field from the Ulster Railway, or from horse-drawn coaches, did not exist. In 1837, giving evidence before the Drummond Commission on Irish railways, James McCleery, Secretary and Registrar to the Company, said: 'the Ulster Railway will not materially, if at all, interfere with the conveyance of goods on the Lagan Navigation'.

38. 36 Vic. *c.* 65 (local and personal).

39. *Report of the Commissioners appointed to inquire respecting the system of Navigation which connects Coleraine, Belfast and Limerick*, H.C. 1882 [C. 3173] XXI, 101: evidence of W. R. Rea, Secretary to the Lagan Navigation Co.

40. *Second Report of the Irish Railway Commission* (1838), op. cit., App. B, Sect. 6: Inland Navigations, pp. 50 *et seq.*

41. H.C. 1882 [C. 3173], XXI, 101 (as 39 above).

42. H.C. 1837–8 (145), XXXV, 449 (as 40 above).

43. The Ulster Canal and Tyrone Navigation Act—a private Act (51 and 52 Vic., 24 July 1888).

44. Mr W. J. Forster of the Northern Ireland Ministry of Commerce permitted me to examine a small ledger formerly belonging to the Lagan Navigation Company. This contained a wealth of statistical and factual data on the Lagan and Tyrone Navigations and the Ulster Canal for the years after the amalgamation

of 1888. From it most of the statistics relating to the later history of the canal have been taken.

45. In the spring of 1962 (Order in Council, 12 March 1962), responsibility for the derelict waterways formerly under the control of the Upper Bann Navigation Trustees and the Lagan Navigation Co. was transferred to the Ministry of Finance. This was a natural step, since Works Division engineers from Finance had since 1954 carried out various drainage and maintenance work as 'agents'.

Notes to Chapter IV (p. 62)

1. J.H.C.I., 20 May 1709, III, p. 426; 28 May 1709, III, p. 441.
2. T. Prior, *A List of Absentees of Ireland: Observations on the present trade and condition of that Kingdom*, Dublin, 1727, pp. 285–7.
3. A. Dobbs, *Essay on the Trade and Improvement of Ireland*, Dublin, 1729, p. 94.
4. Haliday Pamphlets: R.I.A., Vol. 87. This pamphlet gives much information on physical and economic conditions in the early years of coal working in east Tyrone.
5. J.H.C.I., 2 November 1753, IX, p. 94.
6. Hugh Boyd, *Letter to a Member of Parliament on the late scarcity of coals in Dublin*, Dublin, 1749. Linenhall Library, Belfast, N. 2591. W. Harris, *The Antient and Present State of the County of Down*, Dublin 1744. R. Barton, *A Dialogue of some things of importance to Ireland, particularly to the County of Ardmagh*, Dublin, 1751.
7. Anon., *A Letter to a Commissioner of the Inland Navigation concerning the Tyrone Collieries*, Dublin, 1752. 'Publicola', *The Irish Collieries and Canal defended*, Dublin, 1752 (both in Haliday Pamphlets; R.I.A., Vol. 243).
8. Young, *Tour in Ireland* (Hutton ed.), II, p. 124. The expenditure of public money on unsuccessful attempts to establish large-scale collieries was particularly lavish. For a fascinating account of general dishonesty, incompetence and waste, see J. F. Fetherston, *A Narrative of Facts concerning the Tyrone Collieries*, Dublin, 1771. Haliday Pamphlets: R.I.A., Vol. 366.
9. J.H.C.I., 17 Nov. 1767, VIII, Pt. II, p. 179, *Report*.
10. J.H.C.I., 12 Nov. 1761, XII, pp. 482–4.
11. Young, *Tour in Ireland*, op. cit., II, p. 127.
12. J.H.C.I., 17 Nov. 1767, VIII, Pt. II, p. 180, *Estimate*.
13. His name has always proved something of a stumbling block, being spelt Dukart, Ducarte, Ducate, Dularte, Duchart and Ducart. It seems really to have been Daviso de Arcort. His origin is unknown. His will is dated 30 Nov. 1780 and was proved on 29 March 1786, but the date of his death is not known. See also Maurice Craig, *Dublin, 1660–1860*, London, 1952, pp. 134, 169, and Young, *Tour in Ireland*, Hutton ed., II, p. 127.
14. V. I. Tomlinson, 'Early Warehouses on Manchester Waterways', *Trans. Lancs. & Cheshire Ant. Soc.*, Vol. 71, 1961, pp. 129 *et seq.*
15. F. Mullineux, 'The Duke of Bridgewater's Underground Canals at Worsley', *Trans. Lancs. & Cheshire Ant. Soc.*, Vol. 71, 1961, pp. 152 *et seq.*
16. J.H.C.I., 7 May 1768, VIII, Pt. II, pp. 244 *et seq.*
17. For Ducart's inclined planes, see D. H. Tew, 'Canal Lifts and Inclines', *Trans. Newcomen Soc.*, XXVIII, 1951–2, p. 35; A. Rees, *Cyclopaedia*, London, 1819, art. 'Canal'; *Reports of the late John Smeaton*, 1812, II, p. 278.

18. Map by Francis Sloane, 1786, R.I.A., M.R. Recess, Case 1, Shelf A. Measurements from *Report of Richard Owen on the several modes of communication between the Tyrone Navigation and the Collieries of Drumglass*, J.H.C.I., 20 February 1787, XXIV, p. 173.

19. William Chapman, *Canal Navigation*, 1797, gives the dimensions of the boats as 4½ ft wide × 2½ ft high × 10 ft long. He describes them as straight-sided and flat-bottomed, with one end square and the other end pointed, moving along end to end.

20. Rees, *Cyclopaedia*, op. cit.

21. Ibid.

22. The canal is shown as complete on Taylor and Skinner's road atlas of 1778.

23. John Brownrigg's *Report* to the Directors General, 22 October 1801, P.R.O.I., P.W.I., 5/7/1, pp. 321 *et seq.*

24. *Report from the Commissioners appointed to inquire into the state of the Navigation from Lough Neagh to the Collieries*, J.H.C.I., 13 June 1785, XXII, p. 546.

25. *Report of Richard Owen on the several modes of communication between the Tyrone Navigation and the Collieries of Drumglass*, J.H.C.I., 20 February 1787, XXIV, p. 173.

26. J.H.C.I., 16 February 1788, XXV, p. 182.

27. J.H.C.I., 25 May 1789, XXVI, p. 284.

28. *Belfast News-letter*, 16 April 1797.

29. J.H.C.I., 22 December 1786, XXIII, p. 137.

30. W. W. Seward, *Topographica Hibernica*, Dublin, 1795, p. 43.

31. During the closing years of the eighteenth century flooding seems to have been a serious problem. It is therefore interesting that in 1786 'the sum of 2,000 l. was granted to John Staples and James Caufeild, Esqrs., for the purpose of erecting a steam engine at their Collieries in the County of Tyrone' (J.H.C.I., 22 April 1786, XXIII, App., p. cxix). On Francis Sloane's map of 1786, a 'Fire Engine' is marked at the Drumglass Collieries, in the extreme west of the area shown, probably that for which the parliamentary grant had been made earlier in the year. It was one of the earliest steam engines erected in Ireland. An account of its erection is in the Staples collection in the P.R.O.N.I. (D.1567/31), while it is mentioned in a description of the Drumglass Collieries in 1789, from a similar source.

32. Seward, *Topographica Hibernica*, p. 43.

33. A great deal of material for the period of control by the Directors General is available in the Public Record Office Ireland in Dublin. The general reference is P.W. I. (Public Works Ireland).

    Walker's estimate appears in the earliest Letter Book (Northern District), P.W. I. 5/3/1, pp. 96–9.

34. P.W. I 5/3/1, p. 176.

35. For Brownrigg's report on the Newry and Coalisland Canals, see P.W. I 5/7/1, pp. 321 *et seq.*

36. For Monks' report on the Coalisland Canal, see P.W. I 5/7/1, pp. 332 *et seq.*

37. Pemberton's name and the date, 1808, can still be seen on a plaque on the eastern parapet wall of the humped bridge spanning the double lock (Mack's Lock) about 1½ miles south-east of Coalisland. He was also employed on the reconstruction of several of the locks on the inland section of the Newry Navigation, but was dismissed here as early as 1801, following complaints against Henry Walker, the engineer, for alleged faulty workmanship at the Poyntzpass lock.

38. *Thirteenth Report of the Commissioners appointed to inquire into the fees etc. in certain public offices in Ireland: Directors General of Inland Navigation.* H.C. 1812–13 (61), VI, 317.

39. Edward Wakefield, *An Account of Ireland, Statistical and Political*, 1812, 2 vols, I, pp. 638–41.

40. W. McEvoy, *Statistical Survey of County Tyrone*, Dublin, 1802, p. 21.

41. J. Gough, *A Tour in Ireland in 1813 and 1814*, Dublin, n.d., p. 59.

42. In 1812 for example, it was reported (Gough, *Tour in Ireland*, p. 59) that the local fuel was 'small, soft and without bituminous quality to make it adhere', while in 1817 the Honourable Member for County Monaghan abstained from the general condemnation of Tyrone coal so widespread at that time, drawing attention to its great utility: 'every gentleman's residence is liable to accidental fire and a throwing on of those coals is the surest and most effectual way of putting it out'. *Newry Magazine*, III, 1817, p. 140.

43. Manuscript notebooks compiled during the course of the first Ordnance Survey of Ireland, during the 1830s. The relevant parishes for information on the Tyrone Navigation are Clonoe and Tullyniskan (Boxes 51 and 52, R.I.A.).

44. *Second Report of the Irish Railway Commission* (1838), op. cit., App. B, Sec. 6: Inland Navigations, pp. 50 *et seq.*

45. Lewis, *Topographical Dictionary of Ireland*, I, p. 384.

46. The Ballinamore & Ballyconnell Canal between the Shannon and the Woodford River, entering Upper Lough Erne, was built between 1846 and 1860 at a cost of almost £250,000. During its short working life, from 1860 to 1869, it handled eight or nine boats in all. By 1880 it was largely derelict.

47. See chapter VII.

48. D. B. McNeill, *Coastal Passenger Steamers and Inland Navigations in the north of Ireland*, Belfast, 1960, pp. 23–5.

49. (i) MS. notebooks (memoirs) for the Parish of Clonoe (Royal Irish Academy; Box 51). (ii) App. to John McMahon's *Report* of 1846: *Report to the Commissioners appointed under the provisions of 5 and 6 Vic. cap. XXIX on the drainage of the flooded lands and the improvement of the navigations in the Lough Neagh district* (see Chapter VII below). (iii) R. Manning, *Drainage and Navigation in the Lough Neagh district*, Dublin, 1884, pp. 55 *et seq.* (iv) *Report of the Commissioners appointed to inspect the accounts and examine the works of railways in Ireland*, H.C. 1867–8 [4018] XXXII, 469: App., Pt. IV: Canals and River Navigations, p. 176.

50. *Report of the Committee appointed to inquire into the Board of Works, Ireland*, H.C. 1878 [C. 2060] XXIII, I, App. H, No. 3, p. 273.

51. Ibid., p. 174 (evidence of J. G. V. Porter of Lisbellaw, Co. Fermanagh).

52. See Chapter III, note 44.

53. Statistics for 1844 from the Appendix to McMahon's Report of 1846 (see note 49 (ii)), those for 1914 from H. L. Glasgow, *The Upper Bann Navigation Rate*, Cookstown, 1916, p. 10.

54. *Final Report of the Royal Commission appointed to inquire into the Canals and Inland Navigations of the United Kingdom*: XI (Ireland), H.C. 1911 [Cd. 5626], XIII, 19.

55. The journey from Belfast to Ellis' Gut took 2½ days, that to Coalisland 3½, depending upon how long the lighter had to wait at 'The Gut' for a tug to tow across the treacherous waters of southern Lough Neagh. R. A. Scott-James, *An Englishman in Ireland*, London, 1910. Scott-James describes a journey by canoe from Stranmillis to Athlone in the summer of 1910.

56. See Chapter III, note 44.

57. Between 1924 and 1926 these deep workings produced over 36,000 tons, most of which was sent by rail to Belfast. Coal mining ceased abruptly in the latter year, following the discovery of major faulting at depth. W. A. McCutcheon, *The Collieries of east Tyrone from the mid-seventeenth century to the present day*, typescript thesis in the library of the Queen's University of Belfast, 1958.
58. See Chapter III, note 44.

Notes to Chapter V (p. 86)

1. 31 Geo. III *c.* 45 (local and personal). Towards this sum of £11,000, the Irish Parliament granted £3,703 of debentures, bearing interest at 4 per cent, before construction work began.
2. One indenture of particular interest was that concluded between Abercorn and the Bishop of Derry, Frederick, Earl of Bristol. Attached to it was a map of 1792 showing the proposed line of the new canal and the progress already made by October of that year.
3. J.H.C.I., 29 January 1793, 10 February 1794, 14 February 1795, 29 January 1796 and 27 January 1797.
4. *Londonderry Journal*, 5 April 1796.
5. *Strabane Journal*, 28 March 1796.
6. Marmion, *Maritime Ports of Ireland*, p. 413.
7. Lewis, *Topographical Dictionary of Ireland*, II, p. 576. *Parliamentary Gazetteer*, III, pp. 280–1.
8. Lewis, *Topographical Dictionary of Ireland*, II, p. 576.
9. *Thom's Irish Almanac and Official Directory for the year 1849*, Dublin, 1849, pp. 162 *et seq.*
10. *Report of the Commissioners appointed to inspect the accounts and examine the works of railways in Ireland*, H.C. 1867–8 [4018] XXXII, 469: App. Pt. IV: Canals and River Navigations, p. 165.
11. A large collection of material relating to the various companies that controlled the Strabane Canal from 1860 are with Thomas Elliott & Son, Solicitors, Abercorn Square, Strabane, Co. Tyrone; much of this information has been drawn from this source.
12. *Report on the condition of the Strabane Canal*, 10 September 1898. This report, by Lt.-Col. G. W. Addison (Asst. Sec., Railway Dept., B. of T.) was made in compliance with an order of 22 June 1898: Elliott & Son's collection (note 11).
13. *Report on the condition of the Strabane Canal*, 2 December 1905. This report, by Major J. Pringle, R.E., followed further complaints on the dangerous nature of the embankments.
14. E. M. Patterson, *The County Donegal Railways*, 1963.
15. For detailed information on the condition of the waterway at this time and on the extent and nature of the traffic being carried see various reports of the Royal Commission of 1906–11.

Notes to Chapter VI (p. 98)

1. J.H.C.I., 2 July 1800, App., Vol II, pp. mlxxxii–mlxxxiii.
2. *Papers respecting a Survey for a Navigation between Lough Erne and Lough Neagh*, February 1815, National Library of Ireland, Joly Pamphlets, No. 626.
3. Letter Book, Lagan Navigation, 1809–26. P.R.O.N.I., C.O.M. I/21, p. 182.
4. *Belfast News-letter*, 21 May 1822; 14 June 1822; 21 June 1822.
5. 6 Geo. IV *c.* 193 (5 July 1825).
6. Bernard Mullins, *Thoughts on Inland Navigation* etc., Dublin, 1832, pp. 37 *et seq.*
7. Thomas Casebourne, 'Description of a portion of the Works of the Ulster Canal', *Mins. Proc. Inst. C.E.'s. II* (1842–3), pp. 52, 53.
8. Many of the bridges carrying minor roads across the Ulster Canal had to be built to accommodate the junction of roads running on different levels. Before canal construction a minor road had often sloped abruptly down to a major road, joining it at a normal T junction. The Ulster Canal, running closely parallel to and above the main Armagh-Monaghan turnpike for long distances, made such simple junctions impossible. The engineers had therefore to carry the minor road across the canal by a normal bridge to the line of the main road, but some 20 to 30 ft above it, and then build sloped approaches to the roadway and roughly parallel to it. Separating these inclines from the main road was a long, blank stone wall, highest in the centre opposite the canal arch and sloping smoothly down to the two junctions. From their shape these bridges were often referred to as 'ball alley' bridges.
9. M. B. Mullins, 'An Historical Sketch of Engineering in Ireland', *Trans. Inst. of C.E.'s of Ireland*, VI (1859–61), pp. 1–181, esp. pp. 63–6.
10. While the Ulster Canal was being built, the company had in 1837 and 1838 contributed to the cost of surveys for a canal from Upper Lough Erne south-westwards to the Shannon. Work on this, the Ballinamore & Ballyconnell Canal (see p. 110) did not begin till after the Ulster Canal had been completed, in 1846.
11. This line, the Ulster Canal & Union Railway, never materialized, though the land between Armagh and Caledon was surveyed by James Boyle and John Godwin, engineer to the Ulster Railway Company. Killaly's original survey of the Ulster Canal in 1815 included provision for a navigation offshoot towards Armagh. Now it was being proposed to bridge this same gap by rail.
12. Cited in the *Second Report of the Royal Commission on Irish Public Works*. H.C. 1888 [C. 5264] XLVIII, 143.
13. Several maps in the P.R.O.N.I. (C.O.M. I/M108, /M121, M/134) show the engineering work carried out between 1865 and 1873.
14. *Report of the Committee appointed to inquire into the Board of Works, Ireland*. H.C. 1878 [C. 2060] XXIII, 1: introduction, pp. xxxix *et seq.*
15. Ibid, pp. 130–1.
16. These were (a) *Report of the Committee appointed to inquire into the Board of Works, Ireland*. H.C. 1878 [C. 2060] XXIII, 1. (b) *Report of the Commissioners appointed to inquire respecting the system of navigation which connects Coleraine, Belfast and Limerick*. H.C. 1882 [C. 3173] XXI, 101.
17. [C. 3173] 1882. Minutes of evidence Nos. 1562, 1622 and 1690–1748. For further information on Ulster Canal traffic at this time see the records of the Board of Public Works in the P.R.O.I. These include (P.W. II, 5/29/1–11)

registers of boats arriving at and departing from all the principal stations on the canal from 1873 to 1888, with full details of cargoes, tolls, total receipts, destinations, etc.

18. See Chapter III, note 44.
19. Scott-James, *Englishman in Ireland*, pp. 93, 115–16.
20. From about 1896 until the Royal Commission (1906–11) a great many letters passed between the Office of Public Works in Dublin and the Lagan Navigation Co. concerning the Ulster Canal (misc. documents and correspondence with the Ministry of Commerce, Govt. of Northern Ireland, successors to the Lagan Navigation Co.). These were mostly requests by the Belfast company for financial assistance in operating the Ulster Canal, or pleas for the rescinding of the Act of 1888 authorizing the transfer, and refusals by the Office. Many are printed as Appendices to Vol. II, pt. 2 [Cd. 3717] of the Report of this Commission, as is a pamphlet, *The Ulster Canal and Lagan Navigation Company* (Appendix 13), a bitter attack by the Secretary of the Lagan company on the Board of Public Works (see also Appendix No. 4 in Vol. XII: H.C. 1911 [Cd. 5653] XIII, 123).

The recommendation referred to here is contained in the *Final Report of the Royal Commission*, XI (Ireland) H.C. 1911 [Cd. 5626] XIII, 19, p. 25.

Notes to Chapter VII (p. 120)

1. Cited in *Report on Drainage and Navigation in the Lough Neagh district*, by Robert Manning, chief engineer to the Board of Public Works, Dublin, 1884, p. 17: in P.R.O.N.I., F.I.N. IX.
2. Amongst those who later referred to the navigational possibilities of the Lower Bann and Lough Neagh were Sir Charles Coote, the Rev. John Dubourdieu, Richard Owen, Thomas Townsend, William Richardson, Thomas Woodhouse, Sir John Macneill, William Gregory and Thomas Rhodes.
3. Manning, *Report* (see note 1), p. 19.
4. 5 and 6 Vic. *c.* 79 (5 August 1842). The Commissioners of Public Works (Ireland) were appointed 'agents' to ensure the effective operation of this legislation.
5. These Memorials of 1844 and 1845 are in a volume of miscellaneous Acts and Awards, Bye Laws and Regulations, Extracts and Reports, in the possession of Engineers of the Works Division of the Northern Ireland Ministry of Finance.
6. *Belfast News-letter*, 9 January 1844.
7. McMahon's 'Report to the Commissioners appointed under the Provisions of 5 and 6 Vic. cap lxxix on the drainage of the flooded lands and the improvement of the navigation in the Lough Neagh district' which is of basic importance in the subsequent history of inland navigation in Lough Neagh and along the Lower Bann, is printed as an Appendix to Manning's Report of 1884 (op. cit., pp. 12–32).
8. Harding's *Report on the Valuation of the flooded and injured lands in the district* was published with McMahon's Report in 1846.
9. 'Statement of the lands to be drained and improved, including those within the districts or limits within which it is proposed to put the powers of the said Acts into execution, together with the names of the reputed proprietors.' Robert Harding to the Secretary of the Board of Public Works (Ireland), 16 April 1846 (Manning, *Report* (see note 1), pp. 39–46).

L

10. 'Estimate of the cost of the proposed works of Navigation and Schedule show-ing the district likely to be benefited by the improvement of the navigation of the Lower Bann river, Lough Neagh and the navigable rivers flowing into the same . . . and the proportions in which the same are likely to be benefited and should contribute to the proposed works of navigation.' Robert Harding to the Secretary of the Board of Public Works (Ireland) 8 May 1846 (Manning, *Report* (see note 1), pp. 47–52).

11. McMahon report of 1846, Manning, op. cit., p. 55 (*see* notes 1 and 7).

12. On 11 August 1858, a copy of the findings of the Commissioners of Public Works upon the fixing of financial schedules was deposited with the Clerks of the Peace of each County concerned, and published in the local newspapers.

13. The inquiry was well attended; it appears that objections were to the assessment of drainage dues. The navigation assessment went unchallenged and this por-tion of the Award was confirmed and enrolled in Chancery on 13 April 1859. (This Award, dated 18 February 1859, is in a volume in possession of Engineers of the Works Division in the Northern Ireland Ministry of Finance.)

14. McMahon report of 1846, Manning, op. cit., p. 29 (see note 7).

15. Ibid., Manning, op. cit., pp. 29–30.

16. *Report of the Committee appointed to inquire into the Board of Public Works, Ireland,* H.C. 1878 [C. 2060] XXIII, 1, pp. xxxix *et seq.*; also App. A: II, no. 7, p. 271.

17. Glasgow, *The Upper Bann Navigation Rate* (see Chapter IV, note 53), pp. 13–15.

18. [C. 3173] 1882.

19. *First Report of the Royal Commission on Irish Public Works.* H.C. 1887 [C. 5038] XXV, 471, p. 37.

20. 'Bann and Lough Neagh Drainage Report', Sir A. R. Binnie to the Lord Lieutenant, H.C. 1906 [Cd. 2855] XCVI, 901, pp. 17–19.

21. See also F. J. Dick's criticism of Binnie's Report (MS. additions to Cd. 2855 of 1906) in the P.R.O.N.I. Binnie's Report of 1906 was the first *engineer's* report dealing with the problems of flooding in the Lough Neagh basin and the Lower Bann which presupposed the abandoning of the navigation works below Toomebridge, all previous reports having supposed that the navigation would be modified rather than abandoned (see also 'Drainage of Lough Neagh and the Lower Bann—Report' by F. J. Dick: H.C. 1904 [Cd. 2205] LXXIX, 21).

22. *Final Report of the Royal Commission,* XI (Ireland), H.C. 1911 [Cd. 5626] XIII, 193, pp. 28 and 29.

23. 15 and 16 Geo. V, Ch. 24. 'An Act to empower county councils to carry out schemes for the drainage and improvement by drainage of lands, to provide for the transfer to county councils of the functions of certain drainage authorities, to enable further powers to be granted to other existing drainage authorities, and to establish a Drainage Advisory Committee, and for other purposes con-nected therewith.'

24. *Coleraine Chronicle,* 27 February 1926, 5 February 1927.

25. These were (i) Interim Report on the Drainage of Lough Neagh and the River Bann [Cmd. 72] 1927: (ii) Final Report [Cmd. 90] 1928.

26. [Cmd. 72] 1927, pp. 11–13.

27. 20 Geo. V, Ch. 20. 'An Act to constitute the Ministry of Finance as a Drainage Authority for the purpose of the drainage of Lough Neagh and the River Bann and to abolish certain drainage and navigation authorities.'

28. Details of works carried out are contained in the Navigation Award, dated 18 February 1859 (see note 13).

29. The original sum repayable on improvements in navigation on the Upper and Lower Bann, Blackwater and Lough Neagh was £37,137 10s. od. This had been repaid by 31 March 1877, the portion repaid on account of improvements on the Blackwater and Upper Bann amounting to £9,210 2s., repaid in five instalments.
30. 19 and 20 Vic. *c.* 62 (21 July 1856). 'An Act to provide for the maintenance of Navigations made in connexion with Drainage and to make further provision in relation to works of drainage in Ireland.'
31. *Final Report of the Royal Commission,* XI (Ireland), H.C. 1911 [Cd. 5626], XIII, 19, p. 30.
32. Glasgow, *The Upper Bann Navigation Rate* (see Chapter IV, note 53), p. 14.
33. In fact, the work would have been carried out by the Lough Neagh Drainage Trust at the expense of the landowners of a large area north and south of Lough Neagh.
34. Glasgow, *The Upper Bann Navigation Rate* (see Chapter IV, note 53).
35. The timber staging and movable pontoon have now been removed, and the mainland linked to Derrywarragh Island by a permanent causeway.
36. Statistics and information from Account Books and Minute Books of the Upper Bann Navigation Trustees; P.R.O.N.I., C.O.M. I/10/11; C.O.M, I/10/21; C.O.M. I/10/31.
37. Ibid.

Notes to Chapter VIII (p. 147)

1. G. V. Sampson, *Memoir . . . of the County of Londonderry,* 1814, p. 238.
2. *Londonderry Journal,* 19 March 1776, 11 October 1785, 1 December 1807 1 August 1809. See E. M. Patterson, *The Londonderry & Lough Swilly Railway,* 1964, pp. 12–13.
3. P.R.O.N.I.: Massereene-Foster collection. D. 207/27/32.
4. Field Memoirs, Ordnance Survey, County Tyrone, Parish of Aghalurcher, letter from James Spiller, 19 June 1824, to James Sinclair of Hollyhill, Strabane, Co. Tyrone.
5. Field Memoirs, Ordnance Survey, Parish of Cappagh, County Tyrone.

APPENDIX I

Summary of Facts about the Canals and Navigations of the North of Ireland

A. *Rivers and Lakes Successfully Made Navigable*

River	Date of Act under which Work was begun	Date Wholly Opened	Approx. Cost at Opening	Terminal Points	Length including Branches
Lower Bann and Lough Neagh	1847	1859		Toomebridge to Coleraine, and the northern part of Lough Neagh	32 miles 32 chains
Upper Bann and Lough Neagh	1847	1858		Along river Blackwater from Lough Neagh to Blackwatertown, and along Upper Bann from Lough Neagh to White-coat Point. Also the southern part of Lough Neagh	21 miles 28 chains

Greatest Number of Locks	Size of Boats taken	Date of Disuse for Commercial Traffic	Date of Abandonment	Present Ownership
5 (one a double lock)	Could take *c* 120 ft × 19 ft. In 1900 actual craft were *c*. 62 ft × 14 ft 6 in. draught 5 ft 6 in.	Still used through first lock at Toomebridge	Open	Ministry of Finance, Govt. of Northern Ireland
None	Unlimited except in draught In 1906, actual craft *c*. 62 ft × 14 ft 6 in,. draught 5 ft 6 in. (5 ft on Blackwater)	Early 1950s	Abandoned 1954	Ministry of Finance, Govt. of Northern Ireland

B. *Rivers with Uncompleted Navigation Works—None*
C. *Canals, the Main Lines of which were Completed as Authorized*

Canal	Date of Act under which Work was begun	Date Wholly Opened	Approx. Cost at Opening	Terminal Points	Branches Built
Coalisland Canal (Tyrone Navigation)	1729	1787	£31,417	Coalisland—R. Blackwater	
Ducart's Canal		1777	£18,750	Coalisland—Drumglass	
Lagan	1753	First section 1763; second section 1794	First section £43,304; second section £62,000	Belfast (Stranmillis)—Lough Neagh (Ellis' Gut)	—
Newry	1729	1742	£52,000	Newry–Whitecoat Point on Upper Bann	—
Newry Ship	1829	1850	c. £110,000	Canal: Newry–Upper Fathom Navigable channel: Upper Fathom–Warrenpoint	
For the first ship canal at Newry the information is as follows	1755 [29 Geo. II (Irish) c.1.]	1769	£23,000	Newry: townland of Lower Fathom	

Length	Greatest Number of Locks	Size of Boats Taken	Date of Disuse for Commercial Traffic	Date of Abandonment	Present Ownership
4 miles 32 chains	7 (one a double lock)	62 ft × 14 ft 6 in. Draught 5 ft	1946	1954	Ministry of Finance, Govt. of Northern Ireland
3½ miles	3 inclined planes	10 ft × 4½ ft 2½ ft high	By 1787	None	None
25 miles 66 chains	27	62 ft × 14 ft 6 in. Draught 5 ft 6 in	1951	Sprucefield (Lisburn)—Ellis' Gut, 1954; Sprucefield–Stranmillis, 1958	Ministry of Finance, Govt. of Northern Ireland
18½ miles	14 (after 1810, 13)	62 ft × 14 ft 6 in. Draught 5 ft 2 in., originally locks 44 ft × 15 ft 6 in	1938–9	Whitecoat Point—outskirts of Newry, 1949; through Newry, 1956	Newry Port and Harbour Trust
Canal 3 miles 4 chains Channel 4 miles	1	205 ft × 32 ft. Draught 13 ft	Open	Open	Newry Port and Harbour Trust
1 m. 60 ch.	2	The canal itself was 60 ft wide, 12 ft deep and the Lock at Lower Fathom was 130 ft × 22 ft, capable of handling vessels up to 120 tons: dimensions of vessels not known	1829	—	—

Canal	Date of Act under which Work was begun	Date Wholly Opened	Approx. Cost at Opening	Terminal Points	Branches Built
Strabane Canal	1791	1796	£11,858	Strabane–R. Foyle	—
Ulster	1825	1841	£231,000	Charlemont on River Blackwater to Wattle Bridge on River Finn	

Length	Greatest Number of Locks	Size of Boats Taken	Date of Disuse for Commercial Traffic	Date of Abandonment	Present Ownership
miles chains	2	Canal could take *c.* 100 ft × 22 ft. In 1900 actual craft were 70 ft × 15 ft 6 in. Draught 5 ft	1932	Strabane-Dysert 1962, rest open	Strabane & Foyle Navigation Co. Ltd.
5 miles 7 chains	26	62 ft × 11 ft 6 in. Draught 5 ft	1929	1931	—

APPENDIX II

Principal Engineering Works

A. *Inclined Planes*

Canal	Name of Plane	Vertical Rise	Dates Working	Notes
Ducart's	Brackaville Drumreagh Farlough	55 ft 65 ft 70 ft	Intermittently 1777–87	Counterbalanced, with horse gins to get boats over the upper sill

B. *Lifts*

None.

C. *Tunnels over 500 yards*

None.

D. *Outstanding Aqueducts*

Lagan Navigation   Spencers Bridge (Lagan) (Goatum)

Ducart's Canal   Newmills (Torrent)

INDEX